The **BEST WRITING** on **MATHEMATICS**

2014

The **BEST WRITING on MATHEMATICS**

2014

Mircea Pitici, Editor

PRINCETON UNIVERSITY PRESS
PRINCETON AND OXFORD

Copyright © 2015 by Princeton University Press
Published by Princeton University Press, 41 William Street,
Princeton, New Jersey 08540
In the United Kingdom: Princeton University Press,
6 Oxford Street, Woodstock, Oxfordshire OX20 1TW

press.princeton.edu

ISBN (pbk.) 978-0-691-16417-5

This book has been composed in Perpetua

Printed on acid-free paper. ∞

Printed in the United States of America

1 3 5 7 9 10 8 6 4 2

for my daughter Ioana, again

Contents

Introduction
MIRCEA PITICI — ix

Mathematics and the Good Life
STEPHEN POLLARD — 1

The Rise of Big Data: How It's Changing the Way We Think
about the World
KENNETH CUKIER AND VIKTOR MAYER-SCHÖNBERGER — 20

Conway's Wizards
TANYA KHOVANOVA — 33

On Unsettleable Arithmetical Problems
JOHN H. CONWAY — 39

Color illustration section follows page 48

Crinkly Curves
BRIAN HAYES — 49

Why Do We Perceive Logarithmically?
LAV R. VARSHNEY AND JOHN Z. SUN — 64

The Music of Math Games
KEITH DEVLIN — 74

The Fundamental Theorem of Algebra for Artists
BAHMAN KALANTARI AND BRUCE TORRENCE — 87

The Arts—Digitized, Quantified, and Analyzed
NICOLE LAZAR — 96

On the Number of Klein Bottle Types
CARLO H. SÉQUIN — 105

Adventures in Mathematical Knitting
SARAH-MARIE BELCASTRO — 128

The Mathematics of Fountain Design: A Multiple-Centers Activity
MARSHALL GORDON — 144

Food for (Mathematical) Thought
PENELOPE DUNHAM   156

Wondering about Wonder in Mathematics
DOV ZAZKIS AND RINA ZAZKIS   165

The Lesson of Grace in Teaching
FRANCIS EDWARD SU   188

Generic Proving: Reflections on Scope and Method
URI LERON AND ORIT ZASLAVSKY   198

Extreme Proofs I: The Irrationality of $\sqrt{2}$
JOHN H. CONWAY AND JOSEPH SHIPMAN   216

*Stuck in the Middle: Cauchy's Intermediate Value Theorem
and the History of Analytic Rigor*
MICHAEL J. BARANY   228

*Plato, Poincaré, and the Enchanted Dodecahedron: Is the Universe
Shaped Like the Poincaré Homology Sphere?*
LAWRENCE BRENTON   239

*Computing with Real Numbers, from Archimedes to Turing
and Beyond*
MARK BRAVERMAN   251

Chaos at Fifty
ADILSON E. MOTTER AND DAVID K. CAMPBELL   270

Twenty-Five Analogies for Explaining Statistical Concepts
ROBERTO BEHAR, PERE GRIMA, AND LLUÍS
MARCO-ALMAGRO   288

College Admissions and the Stability of Marriage
DAVID GALE AND LLOYD S. SHAPLEY   299

The Beauty of Bounded Gaps
JORDAN ELLENBERG   308

Contributors   315

Notable Writings   325

Acknowledgments   333

Credits   335

Introduction

MIRCEA PITICI

Welcome to reading the fifth anthology in our series of recent writings on mathematics. Almost all the pieces included here were first published during 2013 in periodicals, as book chapters, or online.

Much is written about mathematics these days, and much of it is good, amid a fair amount of uninspired writing. What does and what doesn't qualify as good is sometimes difficult to decide. I have learned that strictly normative criteria for what "good writing" on mathematics should be are of little utility, and in fact, might be counterproductive. Good writing comes in many styles. I know it when I see it, and I might see it differently from the way you see it. Making a selection called "the best" is inevitably subjective and inevitably doubles into an elimination procedure—a combination of successive positive and negative evaluations, always comparative, guided by the preferences, competence, and probity of the people involved, as well as by publishing and time constraints. I have described elsewhere how we reach the table of contents each year, but I find it useful to retell briefly the main steps of the process.

Driven by curiosity and by personal interest, I survey an enormous amount of academic and popular literature. I have done so for decades, even when access to bibliographic sources was much more difficult than it is today. I enjoy pondering what people say, making up my mind about what I read, and making up my mind independently of what I read. When pieces have a chance to be included in our anthology I take note and I put them aside, coming up with many titles each year (see the Notable Writings section at the end of the volume). From this broad collection I leave out pieces too long for the economy of our book or unaffordable because of copyright or reprinting issues. At this stage I usually look at a few dozen articles and I advance them to the publisher.

Princeton University Press asks independent reviewers to read and rate the articles, a procedure meant to guide us further. The reviewers always agree on some pieces but sharply disagree on others. When I consider the reports I also have in mind some goals I envisioned for the series well before I even found a publisher for this enterprise.

I want accessible but nontrivial content that presents for mathematicians and for the general public a wide assortment of informed and insightful perspectives on pure and applied mathematics, on the historical and philosophical issues related to mathematics, on topics related to the learning and the teaching of mathematics, on the practice and practicality of mathematics, on the social and institutional aspects in which mathematics grows and thrives, or on other themes related to mathematics. I aim to offer each volume as an instrument for gaining nondogmatic views of mathematics, impervious to indoctrination—as much as possible inclusive, panoramic, comprehensive, and reflecting a multitude of viewpoints that see through the apodictic nature of mathematics. No doubt some people dread this diversity of viewpoints when thinking about mathematics and adopt a defensive stance, retreating into overspecialization, frightened by opinions that upset their views on mathematics or by the specter of (what they consider to be) dilettantism. For what it is worth, *The Best Writing on Mathematics* series is meant as an antidote to the contagious power that emanates from such fears.

The final content of each book in the series is dependent on the literature available during the latest calendar year; therefore, each volume reflects to some extent the vagaries of fashion and, on the other hand, it is deprived of the topics temporarily out of favor with writers on mathematics. Thus each volume should properly be viewed in conjunction with the other volumes in the series.

Books are more important for what they make us think than for what we read in them. If we manage to destroy humanity but to leave all the books intact, there would be no consolation (except for the surviving worms and ants). Books are chances for the potential unknown of our imagination; their worth consists not only in their literal content but also in what they make us reflect—immediately or later, in accord with or in opposition to what we read. Books on mathematics, and mathematical books proper, have a special place in history. Over the past several centuries mathematical ideas, in conjunction with the enterprising élan of explorers, innovators, tinkerers, and the common folk, have served as catalysts for uncountable discoveries (e.g., geographic, technological, scientific, military, and domestic) and for the expansion of life possibilities, thus contributing to human phenotypic diversity.

At first thought opining on mathematics seems benign, having only an upside; yet it is neither trivial nor free of danger. Writing about mathematics is a form of interpreting mathematics; and interpreting mathematics, whether "elementary" or "advanced" mathematics, is not trifling, even for seasoned mathematicians. Interpreting mathematics leaves room for genuine differences of opinion; it allows you to stand your ground even if everybody else disagrees with you—and it's all about *mathematics*, a thinking domain on which people are supposed to agree, not to dissent. Private interpretation of mathematics is different from doing mathematics; it is a mark of personal worldview. In that sense it can indeed be dangerous and unsettling on occasion, but it is always rewarding as a full-mind activity free of the trappings that plague most institutionalized mathematics instruction (e.g., rote memorizing, repetitive learning, following strict rules, inside-the-box thinking, and dumb standardized assessment). Interpreting mathematics allows individuals to build niches in the social milieu by using unique thinking and acting features. For such a purpose mathematics is an inexhaustible resource, similar to but more encompassing than art. The freedom to interpret mathematics as I please compensates for the constraints inherent in the conventional content accepted by the prevailing communitarian view of mathematics. By interpreting, I recoup the range of imaginative possibilities that I gave up when submitting to the compelling rigor enforced by a chain of mathematical arguments. Interpreting mathematics supports my confidence to act not only on my supposed knowledge but also on my ignorance. It enables me to appropriate an epistemological payoff arrangement better than any other I can imagine, since my ignorance will always dwarf my hopelessly limited knowledge. Thus interpreting mathematics breaks paths toward opportunities but also opens doors to peril.

I learned firsthand, at great expense, that voicing opinions about mathematics is risky. I have seen it; I lived it. Long ago—a recent emigrant, poor and naive but buoyantly optimistic—I voiced some of my thoughts on the use and misuse of mathematics during classes at a leading business school. The displeased reaction to what I considered common-sense observations puzzled and befuddled me. To my dismay

I saw that complying, obeying, and blindly turning in assignments—
that mattered, *that* was expected and appreciated, not hard thinking
about the issues at hand. The more I spoke, the stranger the atmosphere
became. For a while I lingered there surrounded by bristling silence,
polite condescension, and impatient tolerance. The denouement came
when a faculty member asked me why I went to that school if I thought I
was so smart. The message, one of the many ensconced in that interpel-
lation, was unmistakable: If you happen to have some ideas (and made
the dumb error of coming here by borrowing a lot of money), better be
careful how you handle them. The next morning I was out of there, at
a staggering financial loss.

That episode had long-lasting consequences for me, highlighting the
fickle rhetoric of the slogans encouraging free inquiry and open dialogue of ideas. When recurring effects and escalating misadventures
added up over the ensuing years, I became a lot more cautious with
what I said—yet matters kept slipping out of control, from terrible to
worse, to desperate, until I chose not to say much anymore. If speaking
up on mathematics proved to be such a disaster, fortunately I accomplished the next to "best" thing for me—the chance to give exposure
to other people's diverse views on mathematics, in the volumes of *The
Best Writing on Mathematics* series.

In conclusion to this preamble, I intend to include here pieces mostly
about mathematics, not necessarily mathematical writing, although a
bit of mathematical exposition makes its way into each volume. Expository mathematical writing is scrutinized, recognized, and rewarded
by publication in the mathematics journals, consideration for professional awards, republication in extended versions as monographs, and
circulation among the mathematical community. But writings *about*
mathematics are spread wide and thin, in venues less frequented by
mathematicians; my goal is to find what is notable in that literature
and to make it easy for people to read it or at least to find it on their
own, quickly and conveniently. In the books of this series you will
meet a deliberate medley of styles, sources, and perspectives. With
this choice for the content we succeed in placing in bookstores volumes that include authors who previously were known only inside the
mathematical community and in circulating among mathematicians
the names of interesting authors who care about mathematics from
the outside.

Contents of the Volume

In the opening piece of the selection, Stephen Pollard reviews John Dewey's conception of human experience and notes that, in the philosopher's views, mathematics and its practice offer an integrative function that transcends the utilitarian goals commonly bestowed on mathematics by enhancing the minds and the lives of the people who study it.

Kenneth Cukier and Viktor Mayer-Schönberger discuss some of the consequences of the recent explosion in digital information, pointing out the qualitative shift it brings to data use and analysis, and concluding that these trends are radically changing our lives, work, and thinking about the world.

Tanya Khovanova poses and solves an ingenious puzzle invented decades ago by John Conway.

In the next piece John Conway observes that some true arithmetical statements are not provable and gives unsophisticated examples that support his assertion.

Brian Hayes traces the history of space-filling curves and offers surprising applications of the counterintuitive process that leads to their construction.

Lav Varshney and John Sun observe that nonlinear, logarithmic scaling is natural to our perception (but distorted into linear scaling by institutionalized education) and contend that, by reducing estimative errors, perceiving logarithmically confers evolutionary advantages.

Keith Devlin formulates negative and positive criteria for evaluating the instructive or noninstructive qualities of games that purport to help children learn mathematics and concludes that most games currently available do not deliver on the educational side.

Bahman Kalantari and Bruce Torrence describe certain features of the "graph" of a polynomial with a complex variable; then they use this intuitive geometrical guidance to draw the main lines in the chain of arguments that proves that any polynomial can be factored into a number of linear factors equal to its degree.

Nicole Lazar reviews the virtues and the pitfalls of using statistical methods to analyze and to create art.

Carlo Séquin catalogs varieties of geometrical shapes similar to a Klein bottle (a geometrical surface with a single face) and briefly indicates the mathematical elements that can help classify them rigorously.

Also starting with a Klein bottle surface, sarah-marie belcastro tackles issues related to the knitting of mathematical objects and illustrates her comments with photographs of needlework.

Marshall Gordon details the activities he developed for teaching eleventh graders the engineering applications of the mathematical study of quadratic quantities and relates the response these activities elicited from the students, stressing that his "multiple-centered" instruction reaches all students according to their levels of interest.

Penelope Dunham brings food into the classroom, not only to facilitate the intuitive understanding of the mathematical concepts encountered in a variety of undergraduate courses she devised but also to stimulate and reward the students.

Dov and Rina Zazkis find that the wonder part of mathematics resides in *doing* mathematics more than in contemplating mathematical results and present examples that link the nature of wonder in mathematics to encountering surprise, counterintuition, and the unexpected.

Francis Su's blog entry, based on a talk he gave as an awardee, is an impassioned plea for renouncing the infatuation with achievement and performance in education and focusing instead on the irreducible humanity that often goes missing in the interaction between teacher and student.

Uri Leron and Orit Zaslavsky reflect on the strengths, limitations, and educational value of *generic proofs*—that is, mathematical proofs based on an example that serves as a springboard for the main ideas supporting the generalization for all cases represented by the example.

John Conway and Joseph Shipman distinguish among different proofs of the same mathematical result based on proofs' dominant features, then exemplify what they call the *proof space* by comparing six ways of proving that the square root of the number 2 is not a proportion of integers.

Michael Barany shows that peculiar conservative inclinations led Augustin-Louis Cauchy to fuse algebraic and geometric thinking, thus inadvertently igniting the theoretical impulse that led to the rigor of modern mathematics.

Lawrence Brenton tells the long story of the speculation concerning the shape of the universe—and how its many avatars relate to geometrical ideas, old and new.

Mark Braverman explains the structural incompatibility between real numbers, which we represent intuitively as a continuous line, and the discrete nature of calculation achievable by mechanical computations. Then he examines the implications for the study of dynamical systems.

Adilson Motter and David Campbell start off by describing the deterministic mind-set characteristic of the exact sciences until about the middle of the twentieth century and continue by tracing the inroads made over the past half-century by the new chaos paradigm and its role in dynamical systems theory.

Roberto Behar, Pere Grima, and Lluís Marco-Almagro list and comment briefly on analogies they find useful for teaching statistics to students who take an introductory course.

David Gale and Lloyd Shapley's article is the only one in this volume that was first published long ago—although attention to it rekindled recently. Gale and Shapley found a well-determined and optimal solution to the problem of assigning people to institutions under conditions of uncertainty.

Jordan Ellenberg reports on an important theoretical breakthrough, Yitang Zhang's proof of the bounded gaps conjecture, and muses on the consequences it might have for the theory of numbers and for our knowledge of randomness.

More Writings on Mathematics

Writing on mathematics for a nonspecialist audience is now a full-time profession for some authors and a pastime for many others. The market is huge and growing. With the caveat that the following list covers only a small part of the publishing output in this area, here are some books that came to my attention over the past year.

I start by noting a volume meant as an overview of the state of mathematics in the United States and as a benchmark for future policy directions, *The Mathematical Sciences in 2025* written by a committee of the National Research Council (for accurate and complete references, please see the list of Books Mentioned at the end of the Introduction).

A book on some of the most important problems in mathematics and physics is Ian Stewart's *Visions of Infinity*. Other presentations of mathematical topics and/or how they spring up in life can be found in Paul Lockhart's *Measurement*, Göran Grimvall's *Quantify!*, Yutaka

Nishiyama's *The Mysterious Number 6174*, Günter Ziegler's *Do I Count?*, Daniel Tammet's *Thinking in Numbers*, David MacNeil's *Fundamentals of Modern Mathematics*, and the nicely illustrated *Mathematics* edited by Tom Jackson. A pocket-size compendium of mathematical notions and results, briefly stated, described, and illustrated, is *Math in Minutes* by Paul Glendinning. An anthology edited by Gina Kolata, *The New York Times Book of Mathematics*, gathers articles published by the newspaper over more than a century. An instructive etymological tool is *Origins of Mathematical Words* by Anthony Lo Bello. And a skeptical view of the powers of mathematical knowledge is laid out by Noson Yanofsky in *The Outer Limits of Reason*.

More technical books with chapters or parts accessible to nonspecialist readers are *The Tower of Hanoi* by Andreas Hinz and his colleagues, *Configurations from a Graphical Viewpoint* by Tomaž Pisanski and Brigitte Servatius, *Magnificent Mistakes in Mathematics* by Alfred Posamentier and Ingmar Lehmann, *The Joy of Factoring* by Samuel Wagstaff, and even Terence Tao's *Compactness and Contradiction*.

Informal introductions to statistics and probabilities are *Will You Be Alive 10 Years from Now?* by Paul Nahin, *The Basics of Data Literacy* by Michael Bowen and Anthony Bartley, *The Cartoon Introduction to Statistics* by Grady Klein and Alan Dabney, and *Dancing on the Tails of the Bell Curve* edited by Richard Altschuler.

Mathematicians have always been fond of telling their own life stories and the ideas that animated them—or others found that it is worth writing about mathematicians' lives and their ideas. The autobiographical literature is picking up speed, with the late Martin Gardner's *Undiluted Hocus-Pocus*, Benoit Mandelbrot's *The Fractalist*, Edward Frenkel's *Love and Math*, Larry Baggett's *In the Dark on the Sunny Side*, and *An Accidental Statistician* by George Box. Key autobiographical elements are also present in Reuben Hersh's anthology of articles *Experiencing Mathematics*. Phillip Schewe has written Freeman Dyson's story as *Maverick Genius*. A historical biography on Ada Lovelace is James Essinger's *A Female Genius*; others are Dirk van Dalen's *L. E. J. Brouwer* and *Vito Volterra* by Angelo Guerraggio and Giovanni Paoloni. A collection of historical studies is *Robert Recorde,* edited by Gareth Roberts and Fenny Smith. A children's book on Paul Erdős is *The Boy Who Loved Math* by Deborah Heiligman. Two books on the history of mathematics at Harvard, with emphasis on personalities, are *A History in Sum* by Steve Nadis and

Shing-Tung Yau, and *Carnap, Tarski, and Quine at Harvard* by Greg Frost-Arnold. And a history of computing through personalities is *Giants of Computing* by Gerard O'Regan.

Other recent works related to the history of mathematics are *If A, then B* by Michael Shenefelt and Heidi White, *Secret History* by Craig Bauer, *Classic Problems of Probability* by Prakash Gorroochurn, *Number Theory* by John Watkins and, with contemporary echoes, the collection *The Legacy of A. M. Turing* edited by Evandro Aggazi. In the philosophy of mathematics, I mention Mark Colyvan's *An Introduction to the Philosophy of Mathematics*, *The Applicability of Mathematics in Science* by Sorin Bangu, and *Plato's Problem* by Marco Panza and Andrea Sereni. A historical-philosophical study is Simon Duffy's *Deleuze and the History of Mathematics*. Highly interdisciplinary are Arturo Carsetti's *Epistemic Complexity and Knowledge Construction* and Pavel Pudlák's *Logical Foundations of Mathematics and Computational Complexity*. Another popular book on complexity, presenting the P versus NP problem, is *The Golden Ticket* by Lance Fortnow. Two books on the philosophy of mind that touch on mathematical reasoning are *Surfaces and Essences* by Douglas Hofstadter and Emmanuel Sander, and *Intuition Pumps and Other Tools for Thinking* by Daniel C. Dennett.

On mathematics education, too many books to mention here are published every year. For now I note first several addressed mainly to teachers and parents: a delightful little book by Stephen Brown titled *Insights into Mathematical Thought*; then *How Math Works* by Arnell Williams, *Success from the Start* by Rob Wieman and Fran Arbaugh, *Captivate, Activate, and Invigorate the Student Brain in Science and Math, Grades 6–12* by John Almarode and Ann Miller, *Math Power* by Patricia Kenschaft, and *One Equals Zero and Other Mathematical Surprises* by Nitsa Movshovitz-Hadar and John Webb. David Tall has published a wide-ranging study on *How Humans Learn to Think Mathematically*. Ed Dubinsky's APOS theory is presented in detail by a group of authors, including Dubinsky but also Ilana Arnon and others. An excellent anthology pieced from the 75 years of National Council of Teachers of Mathematics yearbooks was edited by Francis Fennell and William Speer, with the title *Defining Mathematics Education*. Other collective volumes are *Mathematics & Mathematics Education* edited by Michael Fried and Tommy Dreyfus, *Third International Handbook of Mathematics Education* edited by Ken Clements and his collaborators, *Reconceptualizing Early Mathematics Learning* edited by Lyn English and Joanne Mulligan, *Vital Directions for Mathematics Education Research* edited by Keith Leatham, and *Cognitive Activation in the Mathematics Classroom and Professional Competence of Teachers* edited by Mareike Kunter et al. A book of problems, mainly aimed at preparing students for mathematical competitions, is *Straight from the Book*, by Titu Andreescu and Gabriel Dospinescu.

Accessible books on mathematics and other disciplines or applied to various practical or entertainment endeavors are *Six Sources of Collapse* by Charles Hadlock, *Mathematical Morphology in Geomorphology and GISci* by Daya Sagar, *Mathematical Card Magic* by Colm Mulcahy, and *All the Right Angles* by Joel Levy. An algorithmic fusing of evolutionary, learning, and intelligence perspectives situated in the living environment of organisms is presented by Leslie Valiant in *Probably Approximately Correct*. A decoding and explaining of mathematical motifs in a popular TV series is given by Simon Singh in *The Simpsons and Their Mathematical Secrets*. Readers interested in connections between mathematics and music might find it useful to consult *The Geometry of Musical Rhythm* by Godfried Toussaint and *Mathematics and Music* by James Walker and Gary Don. Those passionate about using mathematics to make origami will find well-illustrated treasure troves of ideas in *Origami Tessellations* by Eric Gjerde and especially the massive *Origami Design Secrets* by the master of the trade, Robert Lang. A more general introduction to the mathematical motifs in the arts is *Manifold Mirrors* by Felipe Cucker. A collection of legal cases and situations that unveil the dangerous mishaps that occur in the (ab)use of mathematics in courtrooms is presented by Leila Schneps and Coralie Colmez in *Math on Trial*. Finally in this category, harnessing mathematics to the search for extraterrestrial life is the leitmotif in *Science, SETI, and Mathematics* by Carl DeVito.

⚬

Every year I include a few online resources, more or less randomly, as I notice them or other people point them out to me.

Philipp Legner is the creator of Mathigon (http://mathigon.org/), an attractive, colorful website where he uses new technologies to build realistic 3-D images of many geometric objects. Another page with good illustrations of solid geometry objects is Paper Models of Polyhedra (http://www.korthalsaltes.com/). A site hosting many images, applets, and videos illustrative of mathematical concepts, from elementary math to multivariable calculus, is Math Insight (http://mathinsight.org/about /mathinsight), maintained by a group of faculty from the University of Minnesota. Useful educational materials at the secondary school level can be found on the Basic Mathematics page (http://www.basic-mathematics .com/) and Interactive Math (http://www.mathsteacher.com.au/index .html); more advanced material is on the Art of Mathematics site (http:// www.artofmathematics.org). AwesomeMath (https://www.awesome math.org/) is a summer program at three universities geared toward gifted secondary students who aspire to receive training beyond school curricula and eventually qualify for advanced mathematical competi- tions. A web page constructed with information taken from Florian Cajori's book *A History of Mathematical Notations* is Earliest Uses of Vari- ous Mathematical Symbols (http://jeff560.tripod.com/mathsym.html). Nate Silver's online magazine FiveThirtyEight (http://fivethirtyeight .com/) posts critical articles on journalistic punditry invoking data and on other public uses of statistics. An online store specializing in math- ematics manipulables is MathsGear (http://mathsgear.co.uk/).

Lastly, the recently restructured site of the Simmons Foundation, under the name *Quanta Magazine* (https://www.simonsfoundation.org /quanta/), hosts some of the best writing on mathematics; I did not include any of the *Quanta* pieces in this volume solely because they are freely and easily accessible online.

〜

I encourage you to send comments, suggestions, and materials I might consider for future volumes to Mircea Pitici, P.O. Box 4671, Ithaca, NY 14852, or send electronic correspondence to mip7@cornell.edu; my Twitter handle is @MPitici.

Books Mentioned

Aggazi, Evandro. (Ed.) *The Legacy of A. M. Turing*. Milano, Italy: Franco Angeli, 2013.

Almarode, John, and Ann M. Miller. *Captivate, Activate, and Invigorate the Student Brain in Sci- ence and Math, Grades 6–12*. Thousand Oaks, CA: Corwin, 2013.

Altschuler, Richard. (Ed.) *Dancing on the Tails of the Bell Curve: Readings on the Joy and Power of Statistics*. Los Angeles: Gordian Knot Books, 2013.

Andreescu, Titu, and Gabriel Dospinescu. *Straight from the Book*. Dallas: XYZ Press, 2012.

Arnon, Ilana, et al. *APOS Theory: A Framework for Research and Curriculum Development in Math- ematics Education*. New York: Springer-Verlag, 2013.

Baggett, Larry W. *In the Dark on the Sunny Side: A Memoir of an Out-of-Sight Mathematician.* Washington, DC: Mathematical Association of America, 2012.

Bangu, Sorin. *The Applicability of Mathematics in Science: Indispensability and Ontology.* London: Palgrave Macmillan, 2012.

Bauer, Craig P. *Secret History: The Story of Cryptology.* Boca Raton, FL: CRC Press, 2013.

Bowen, Michael, and Anthony Bartley. *The Basics of Data Literacy: Helping Your Students (And You!) Make Sense of Data.* Arlington, VA: National Science Teachers Association, 2013.

Box, George E. P. *An Accidental Statistician: The Life and Memories of George E. P. Box.* Hoboken, NJ: Wiley, 2013.

Brown, Stephen I. *Insights into Mathematical Thought: Excursions with Distributivity.* Reston, VA: National Council of Teachers of Mathematics, 2013.

Carsetti, Arturo. *Epistemic Complexity and Knowledge Construction: Morphogenesis, Symbol Dynamics and Beyond.* Heidelberg, Germany: Springer Verlag, 2013.

Clements, M. A. (Ken), et al. (Eds.) *Third International Handbook of Mathematics Education.* New York, NY: Springer-Verlag, 2013.

Colyvan, Mark. *An Introduction to the Philosophy of Mathematics.* Cambridge, UK: Cambridge University Press, 2012.

Cucker, Felipe. *Manifold Mirrors: The Crossing Paths of the Arts and Mathematics.* Cambridge, UK: Cambridge University Press, 2013.

Dennett, Daniel C. *Intuition Pumps and Other Tools for Thinking.* New York: W. W. Norton, 2013.

DeVito, Carl L. *Science, SETI, and Mathematics.* New York: Berghahn, 2013.

Duffy, Simon B. *Deleuze and the History of Mathematics: In Defense of the "New."* London: Bloomsbury, 2013.

English, Lyn D., and Joanne T. Mulligan. (Eds.) *Reconceptualizing Early Mathematics Learning.* New York: Springer-Verlag, 2013.

Essinger, James. *A Female Genius: How Ada Lovelace, Lord Byron's Daughter, Started the Computer Age.* London: Gibson Square, 2014.

Fennell, Francis (Skip), and William R. Speer. (Eds.) *Defining Mathematics Education: Presidential Yearbook Selections, 1926–2012.* Reston, VA: National Council of Teachers of Mathematics, 2013.

Fortnow, Lance. *The Golden Ticket: P, NP, and the Search for the Impossible.* Princeton, NJ: Princeton University Press, 2013.

Frenkel, Edward. *Love and Math: The Heart of Hidden Reality.* New York: Basic Books, 2013.

Fried, Michael N., and Tommy Dreyfus. (Eds.) *Mathematics & Mathematics Education: Searching for Common Ground.* New York: Springer-Verlag, 2014.

Frost-Arnold, Greg. *Carnap, Tarski, and Quine at Harvard: Conversations on Logic, Mathematics, and Science.* Chicago: Open Court, 2013.

Gardner, Martin. *Undiluted Hocus-Pocus: The Autobiography of Martin Gardner.* Princeton, NJ: Princeton University Press, 2013.

Gjerde, Eric. *Origami Tessellations: Awe-Inspiring Geometric Design.* Boca Raton, FL: CRC Press, 2013.

Glendinning, Paul. *Math in Minutes: 200 Key Concepts Explained in an Instant.* New York: Quercus, 2013.

Gorroochurn, Prakash. *Classic Problems of Probability.* Hoboken, NJ: Wiley, 2013.

Grimvall, Göran. *Quantify! A Crash Course in Smart Thinking.* Baltimore: Johns Hopkins University Press, 2010.

Guerraggio, Angelo, and Giovanni Paoloni. *Vito Volterra.* New York: Springer-Verlag, 2013.

Hadlock, Charles R. *Six Sources of Collapse: A Mathematician's Perspective on How Things Can Fall Apart in the Blink of an Eye*. Washington, DC: Mathematical Association of America, 2013.

Heiligman, Deborah. *The Boy Who Loved Math: The Improbable Life of Paul Erdős*. New York: Roaring Brook Press, 2013.

Hersh, Reuben. *Experiencing Mathematics: What Do We Do, When We Do Mathematics?* Providence, RI: American Mathematical Society, 2014.

Hinz, Andreas M., et al. *The Tower of Hanoi—Myths and Maths*. Heidelberg, Germany: Birkhäuser, 2013.

Hofstadter, Douglas, and Emmanuel Sander. *Surfaces and Essences: Analogy as the Fuel and Fire of Thinking*. New York: Basic Books, 2013.

Jackson, Tom. (Ed.) *Mathematics: An Illustrated History of Numbers*. New York: Shelter Harbor Press, 2012.

Kenschaft, Patricia Clark. *Math Power: How to Help Your Child Love Math Even If You Don't*. Revised edition. Mineola, NY: Dover Publications, 2013.

Klein, Grady, and Alan Dabney. *The Cartoon Introduction to Statistics*. New York: Farrar, Straus, and Giroux, 2013.

Kolata, Gina. (Ed.) *The New York Times Book of Mathematics: More than 100 Years of Writing by the Numbers*. New York: Sterling, 2013.

Kunter, Mareike, et al. (Eds.) *Cognitive Activation in the Mathematics Classroom and Professional Competence of Teachers: Results from the COACTIV Project*. New York: Springer-Verlag, 2013.

Lang, Robert J. *Origami Design Secrets: Mathematical Methods for an Ancient Art*. Boca Raton, FL: CRC Press, 2013.

Leatham, Keith R. (Ed.) *Vital Directions for Mathematics Education Research*. New York: Springer-Verlag, 2013.

Levy, Joel. *All the Right Angles—From Gear Ratios to Calculating Odds: Mathematics in the World of Sports*. Buffalo, NY: Firefly Books, 2013.

Lo Bello, Anthony. *Origins of Mathematical Words: A Comprehensive Dictionary of Latin, Greek, and Arabic Roots*. Baltimore: Johns Hopkins University Press, 2013.

Lockhart, Paul. *Measurement*. Cambridge, MA: Belknap Press, 2012.

MacNeil, David B. *Fundamentals of Modern Mathematics: A Practical Review*. Mineola, NY: Dover Publications, 2013.

Mandelbrot, Benoit B. *The Fractalist: Memoir of a Scientific Maverick*. New York: Vintage Books, 2013.

Movshovitz-Hadar, Nitsa, and John Webb. *One Equals Zero and Other Mathematical Surprises:Paradoxes, Fallacies, and Mind Bogglers*. Reston, VA: National Council of Teachers of Mathematics, 2013.

Mulcahy, Colm. *Mathematical Card Magic: Fifty-Two New Effects*. Boca Raton, FL: CRC Press, 2013.

Nadis, Steve, and Shing-Tung Yau. *A History in Sum: 150 Years of Mathematics at Harvard (1825–1975)*. Cambridge, MA: Harvard University Press, 2013.

Nahin, Paul J. *Will You Be Alive 10 Years from Now? And Numerous Other Curious Questions in Probability*. Princeton, NJ: Princeton University Press, 2013.

National Research Council. *The Mathematical Sciences in 2025*. Washington, DC: National Academies Press, 2013.

Nishiyama, Yutaka. *The Mysterious Number 6174: One of 30 Amazing Mathematical Topics in Daily Life*. Kyoto, Japan: Gendai Sugakusha, 2013.

O'Regan, Gerard. *Giants of Computing: A Compendium of Select, Pivotal Pioneers*. New York: Springer-Verlag, 2013.

Panza, Marco, and Andrea Sereni. *Plato's Problem: An Introduction to Mathematical Platonism.* New York: Palgrave MacMillan, 2013.

Pisanski, Tomaž, and Brigitte Servatius. *Configurations from a Graphical Viewpoint.* Heidelberg, Germany: Birkhäuser, 2013.

Posamentier, Alfred S. Ingmar Lehmann. *Magnificent Mistakes in Mathematics.* Amherst, NY: Prometheus Books, 2013.

Pudlák, Pavel. *Logical Foundations of Mathematics and Computational Complexity: A Gentle Introduction.* New York: Springer-Verlag, 2013.

Roberts, Gareth, and Fenny Smith. (Eds.) *Robert Recorde: The Life and Times of a Tudor Mathematician.* Cardiff, UK: University of Wales Press, 2012.

Sagar, B. S. Daya. *Mathematical Morphology in Geomorphology and GISci.* Boca Raton, FL: CRC Press, 2013.

Schewe, Phillip F. *Maverick Genius: The Pioneering Odyssey of Freeman Dyson.* New York: St. Martin's Press, 2013.

Schneps, Leila, and Coralie Colmez. *Math on Trial: How Numbers Get Used and Abused in the Courtroom.* New York: Basic Books, 2013.

Shenefelt, Michael, and Heidi White. *If A, then B: How the World Discovered Logic.* New York: Columbia University Press, 2013.

Singh, Simon. *The Simpsons and Their Mathematical Secrets.* New York: Bloomsbury, 2013.

Stewart, Ian. *Visions of Infinity: The Great Mathematical Problems.* New York: Basic Books, 2013.

Tall, David. *How Humans Learn to Think Mathematically: Exploring the Three Worlds of Mathematics.* Cambridge, UK: Cambridge University Press, 2013.

Tammet, Daniel. *Thinking in Numbers: On Life, Love, Meaning, and Math.* New York: Little, Brown and Co., 2013.

Tao, Terence. *Compactness and Contradiction.* Providence, RI: American Mathematical Society, 2013.

Toussaint, Godfried T. *The Geometry of Musical Rhythm: What Makes a "Good" Rhythm Good?* Boca Raton, FL: CRC Press, 2013.

Valiant, Leslie. *Probably Approximately Correct: Nature's Algorithms for Learning and Prospering in a Complex World.* New York: Basic Books, 2013.

van Dalen, Dirk. *L. E. J. Brouwer: Topologist, Intuitionist, Philosopher: How Mathematics Is Rooted in Life.* New York: Springer-Verlag, 2013.

Wagstaff, Samuel S., Jr. *The Joy of Factoring.* Providence, RI: American Mathematical Society, 2013.

Walker, James S., and Gary W. Don. *Mathematics and Music: Composition, Perception, and Performance.* Boca Raton, FL: CRC Press, 2013.

Watkins, John J. *Number Theory: A Historical Approach.* Princeton, NJ: Princeton University Press, 2014.

Wieman, Rob, and Fran Arbaugh. *Success from the Start: Your First Years Teaching Secondary Mathematics.* Reston, VA: National Council of Teachers of Mathematics, 2013.

Williams, G. Arnell. *How Math Works: A Guide to Grade School Arithmetic for Parents and Teachers.* Plymouth, UK: Rowman and Littlefield, 2013.

Yanofsky, Noson S. *The Outer Limits of Reason: What Science, Mathematics, and Logic Cannot Tell Us.* Cambridge, MA: MIT Press, 2013.

Ziegler, Günter M. *Do I Count? Stories from Mathematics.* Boca Raton, FL: CRC Press, 2014.

The **BEST**
WRITING on
MATHEMATICS

2014

Mathematics and the Good Life

STEPHEN POLLARD

Introduction

A full account of mathematics will identify the distinctive contributions mathematics makes to our success as individuals and as a species. That, in turn, requires us to reflect both on what mathematicians do and on what it means for humanity to flourish. After all, we cannot say how mathematics contributes to our success unless we have a good sense of what mathematicians accomplish and a good idea of what it means for us to succeed. To give our discussion some focus, I offer the following proposition:

> Mathematics makes one substantial contribution to human prosperity: it enhances our instrumental control over physical and social forces.

I say "offer," not "endorse." In fact, I want to persuade you that this proposition is wrong because it presupposes either too narrow a conception of mathematical activity or too narrow a conception of human success. Mathematicians are not just "devices for turning coffee into theorems" (as Erdős may or may not have said). Furthermore, if they *were* such devices, humanity would be the worse for it, and this would be so even if the caffeine-fueled theorem mills were more efficient than real mathematicians at disgorging instrumentally useful product.[1]

The theorem-mill model leaves out at least two vital features of mathematical activity: mathematicians achieve deep insights and have intricately meaningful experiences. These features are vital because they are intrinsic goods for human beings: they are characteristically human ways to prosper. This is not news. It is an ancient idea: mathematics provides insights and experiences that, in themselves, ennoble our species. This idea is one of those enduring themes that make the history of philosophy a long, long conversation rather than a series of disconnected episodes. It can be enervating to feed on the sere remains of antique harvests. It can be energizing, though, to advance a conversation begun of old. We hope to do the latter, with John Dewey as our main interlocutor. We begin, however, with some inspiring words from Bertrand Russell that should make vivid the point of view I am promoting and give you some idea of how I mean to promote it.

A Glorious Torment

In the penultimate chapter of *Education and the Good Life*, Russell considers "the functions of universities in the life of mankind" [1926, p. 311]. He assumes that universities exist for two purposes: "on the one hand, to train men and women for certain professions; on the other hand, to pursue learning and research without regard to immediate utility" [Russell 1926, p. 306]. It is the latter theme, research for the sake of seeking and knowing, that causes Russell's language to soar.[2]

> I should not wish the poet, the painter, the composer or the mathematician to be preoccupied with some remote effect of his activities in the world of practice. He should be occupied, rather, in the pursuit of a vision, in capturing and giving permanence to something which he has first seen dimly for a moment, which he has loved with such ardour that the joys of this world have grown pale by comparison. All great art and all great science springs from the passionate desire to embody what was at first an unsubstantial phantom, a beckoning beauty luring men away from safety and ease to a glorious torment. The men in whom this passion exists must not be fettered by the shackles of a utilitarian philosophy, for to their ardour we owe all that makes man great. [Russell 1926, pp. 312–313]

There is a utilitarian reading even of this passage, a reading that Russell himself endorses. The utilitarian pursuit of knowledge, the quest for greater control over physical and social forces, is not "self-sustaining"; it needs to be "fructified by disinterested investigation" [Russell 1926, p. 312]. Our quest for control profits from seekers who take little or no interest in that quest. Our legitimate utilitarian interests can be served by people who do not give a fig for technological prowess, precisely *because* they do not give a fig.[3] Yes, this is one reason Russell would not fetter mathematicians with the shackles of a utilitarian philosophy. But it is not the only reason. For "even if some splendid theory never has any practical use, it remains of value on its own account; for the understanding of the world is one of the ultimate goods" [Russell 1926, p. 312].[4] This is one half of the view I am trying to promote: the half that emphasizes the intrinsic value of certain *products* of inquiry, the insights that crown a successful inquiry. The other half emphasizes the *process*, reminding us that the whole drama of seeking and finding can be intrinsically good. The hunt itself is one of the glories of human life. It is, to use Russell's phrase, an ultimate good.

Now why might you agree that this is so? Russell's discussion of ultimate goods begins with an argument whose conclusion is not that some particular goods are ultimate, but that there must be ultimate goods.[5]

> The essence of what is "useful" is that it ministers to some result which is not merely useful . . . life must be capable of some intrinsic value: if life were merely useful as a means to other life, it would not be useful at all . . . Somewhere we must get beyond the chain of successive utilities, and find a peg from which the chain is to hang; if not there is no real usefulness in any link of the chain. [Russell 1926, p. 21]

And what are the ultimate goods for human beings? Here Russell offers, not argument, but description. He *shows* us, in arresting word-pictures, admirable ways for human beings to live, making vivid the features he finds intrinsically valuable and inviting us to find them so as well. Here is a striking example from his discussion of how we should present human history to the young.[6]

> I think we should keep in our own minds, as a guiding thread, the conception . . . of the human race as a whole, fighting against chaos without and darkness within, the little tiny lamp of reason growing gradually into a great light by which the night is dispelled. The divisions between races, nations and creeds should be treated as follies, distracting us in the battle against Chaos and Old Night, which is our one truly human activity. [Russell 1926, p. 267]

If we find this passage evocative, if we feel tempted to declaim it "in tones vibrant with manly pathos," we might remind ourselves that "The intrusion of emotion and sentimentality is always the mark of a bad case" [Russell 1932a, pp. 109–110]. Then again, *bad* case is not quite right. There is *no* case in the sense of an argument. There is a depiction meant to be suggestive. If it strikes us in the way it was intended, if we find ourselves more strongly inclined to value things Russell valued in the way Russell valued them (that is, precisely, without benefit of argument, but also without harm to our critical faculties), then the depiction was successful. Can we really hope to do much better?

Russell and Dewey think not. Russell makes his position particularly clear: "Every attempt to persuade people that something is good (or bad) in itself, and not merely in its effects, depends upon the art of rousing feelings, not upon an appeal to evidence" [Russell 1935, p. 235]. This is so because questions of value "cannot be intellectually decided at all" [Russell 1935, p. 243]. Dewey would not be comfortable with this formulation since he himself has no trouble detecting an intellectual element in the determination of values. Extreme emotivism is no better than "oversimplified rationalism": it is wrong to deny "any efficacy whatever to ideas, to intelligence" in the development and propagation of ultimate purposes [Dewey 1939b, p. 150].[7] However, Dewey agrees with Russell that pure ratiocination is unlikely to play the leading role, being much better suited to serve as an adjunct to what Russell calls "preaching." When Russell offers poetry rather than proof, he is, according to Dewey, contributing to the moral progress of our species in the way that has the best chance of success. The "chief instrument of the good" is imagination, not argumentation [Dewey 1934a, p. 348]. "Only imaginative vision elicits the possibilities that are interwoven within the texture of the actual" and, hence, "the sum total of the effect of all reflective treatises on morals is insignificant in comparison with the influence of architecture, novel, drama, on life" [Dewey 1934a, p. 345]. Indeed, "Art has been the means of keeping alive the sense of purposes that outrun evidence and of meanings that transcend indurated habit" [Dewey 1934a, p. 348].[8] Philosophy, when merely prosaic, *can* contribute to moral progress by giving verbal form to important currents in aesthetic experience and providing them an "intellectual base" [Dewey 1934a, p. 345].[9] Make no mistake; a prophetic work of art is, in itself, an *intellectual* achievement of the highest order [Dewey 1934a, pp. 46, 73–74; 1931, p. 116; 1960, p. 198]. Art does not wait upon philosophical prose to draw it into the arena of thought. However, to make our lives whole and to increase our intelligent reflective control over our lives, we need to ponder and discuss human purposes and aspirations when we are not in the grip of aesthetic experiences. Discursive philosophy provides the intellectual base for *that* sort of intelligent reflection and inquiry. Dewey himself provides just such a base with his notion of a "balanced experience" whose cultivation is "the essence of morals" [Dewey 1916, p. 417].

To return for a moment to our main interest, we would like Dewey to teach us about mathematics. We might be expected, then, to focus on his remarks about mathematics. In fact, we largely ignore that material.[10] My slantwise strategy is to focus on a notion central to Dewey's aesthetics and philosophy of education: the aforesaid notion of experience.[11] This notion lies at the heart of Dewey's comprehensive vision of human prosperity and so guides us to a broader understanding of how mathematics helps us flourish as human beings. We now consider human experience in general and, more narrowly, the sort of experiences provided by mathematical inquiry.[12]

Mathematics as Experience

I would like to serve on a jury. I think it would be a good experience. Yes, that would be quite an experience. *An* experience with a definite beginning, a significant internal structure, and an end that is a completion, not just a termination. Experiences, in just this everyday sense, are the building blocks of a human life that is more than an animal existence, a life distinguished by the development and exercise of distinctively human capacities.

The basic conditions of an experience are not uniquely human. "No creature lives merely under its skin" [Dewey 1934a, p. 13]. We animals are all active in an environment that poses problems. How do I escape from the tiger? Where can I find water? Is the defendant guilty? We respond to those problems. If things work out well, we survive, we drink, we learn, we grow. There is a rhythm of problem, activity, resolution, growth in the lives of mollusks and mathematicians.

What is distinctively human is the conscious, reflective, intelligent integration of experiences into a life we can affirm as worthwhile, in which we exercise greater and greater control over a richer and richer array of experiences. If our effort at integration fails spectacularly enough, we will not even be sane [Dewey 1963, p. 44]. If we are unreflective, if we squander opportunities for growth, we will be not only stunted but unfree, "at the mercy of impulses into whose formation intelligent judgment has not entered" [Dewey 1963, p. 65].

So the quality of our experiences and the way those experiences build on one another matters desperately. Why is democracy better than tyranny? Why is kindness better than cruelty? There is, ultimately, a single reason: a democratic arrangement of society and a kindly attitude toward our fellows promote "a higher quality of experience on the part of a greater number" [Dewey 1963, p. 34].[13] Why should we care about the artworks of distant times and places? Because "the art characteristic of a civilization is the means for entering sympathetically into the deepest elements in the experience of remote and foreign civilizations" and, hence, their arts "effect a broadening and deepening of our own experience" [Dewey 1934a, p. 332]. What is education? "It is that reconstruction or reorganization of experience which adds to the meaning of experience, and which increases ability to direct the course of subsequent experience" [Dewey 1916, pp. 89–90].[14]

Experience, experience, experience. Why keep going on and on about experience when we are supposed to be talking about mathematics? Two reasons. First, if we take at all seriously a view that places experience at the heart of human prosperity, we have good reason to reflect on the quality of mathematical experiences and the capacity of those experiences to enrich subsequent experiences. Second, when we do so reflect, we better appreciate how mathematics contributes to human happiness in ways that are not narrowly instrumental. Here are five properties of mathematical experience we consider in turn.

1. Mathematics provides examples of "total integral experiences" [Dewey 1934a, p. 37; 1960, p. 153] of a particularly pure form.
2. Those experiences yield products that allow others to recreate the experiences.
3. Mathematics provides prime examples of experiences that "live fruitfully and creatively in subsequent experiences" [Dewey 1963, p. 28].
4. There is no necessary incompatibility between the ordered growth of mathematical experiences and other desirable forms of human growth.

5. Mathematical experiences not only allow for breadth of cultivation, but once such breadth is achieved, fit easily into the
texture of a variegated life.

We start at the top of the list: the inner structure of a mathematical
experience.

EXPERIENCE

Chimpanzees negotiate environments that offer problems and opportunities. A chimp finds a nut that hides its tasty meat in a hard shell.
The chimp finds a stone. The chimp thoughtfully manipulates the stone
and the nut. Stone cracks shell. Here "the material experienced runs
its course to fulfillment . . . a problem receives its solution . . . a situation . . . is so rounded out that its close is a consummation and not
a cessation" [Dewey 1934a, p. 35; 1960, p. 151]. Whether the chimpanzee is a pioneer or just an apprentice, there is growth. "Life grows
when a temporary falling out [damn this shell!] is a transition to a more
extensive balance of the energies of the organism with those of the
conditions under which it lives" [Dewey 1934a, p. 14]. As with the
chimpanzee, so with the mathematician.

> . . . every experience is the result of interaction between a live
> creature and some aspect of the world in which it lives. . . . The
> creature operating may be a thinker in his study and the environ-
> ment with which he interacts may consist of ideas instead of a
> stone. But interaction of the two constitutes the total experience
> that is had, and the close which completes it is the institution of a
> felt harmony. [Dewey 1934a, pp. 43–44; 1960, p. 160]

The drama of experience runs its characteristic course: problem, activity (physical or mental), success, growth. The problem may be a nut on
the forest floor, an exercise in a book, or a "beckoning beauty" in some
corner of your mind. Chimp or human, if you have any semblance of
puzzle drive (to borrow Feynman's phrase), you are hooked. You *have* to
figure it out. And when the nut is especially hard to crack, when it draws
deeply on your resources, success is graced with an exultation that probably flows from somewhere deep in our primate nature. "Few joys,"
says Russell, "are so pure or so useful as this" [Russell 1926, p. 259]. As
George Pólya observes, even minor triumphs have their savor.

Your problem may be modest; but if it challenges your curiosity and brings into play your inventive faculties, and if you solve it by your own means, you may experience the tension and enjoy the triumph of discovery. Such experiences at a susceptible age may create a taste for mental work and leave their imprint on mind and character for a lifetime. [Pólya 1957, p. v]

Again, a mathematical experience is no mean thing. Though Dewey does not command Russell's and Pólya's insider view of mathematics, he too appreciates this point.

There are absorbing inquiries and speculations which a scientific man and philosopher will recall as "experiences" in the emphatic sense. In final import they are intellectual. But in their actual occurrence they were emotional as well; they were purposive and volitional . . . No thinker can ply his occupation save as he is lured and rewarded by total integral experiences that are intrinsically worthwhile. Without them he would never know what it is really to think and would be completely at a loss in distinguishing real thought from the spurious article. [Dewey 1934a, p. 37; 1960, p. 153]

To see the nut as a call to action, to work at the nut with mind, body, feeling, and purpose, to crack the nut and relish that crack in thought and feeling as a consummation of satisfying, goal-oriented, unconstrained action: that can be a peak primate experience. No wonder, then, that Dewey thinks "science should be taught so as to be an end in itself in the lives of students—something worthwhile on account of its own unique intrinsic contribution to the experience of life" [Dewey 1916, p. 282]. If we had to justify our mathematical preoccupations, we would make a good start by pointing to the character of creative mathematical experiences. But there is more.

RE-CREATION

Dewey observes that "Mathematics and formal logic . . . mark highly specialized branches of intellectual inquiry, whose working principles are very similar to those of works of fine art" [Dewey 1929b, p. 160]. One of the expectations artists and mathematicians share, one working principle, is that their creative experiences leave behind deposits.

Artists deposit novels, paintings, operas. Mathematicians deposit axioms, algorithms, proofs. These deposits are more than memorials of the creator's past activity (as Aristotle and Hegel emphasize). Neither artists nor mathematicians are jealous guardians of "some private, secret, and illicit mode of union with the eternal powers" [Dewey 1929a, p. 34]. Their creations both stimulate and guide *re-creation*. In both mathematics and art, "receptivity is not passivity" [Dewey 1934a, p. 52; 1960, pp. 169–170].

> . . . a beholder must *create* his own experience. And his creation must include relations comparable to those which the original producer underwent . . . there must be an ordering of the elements of the whole that is in form, though not in details, the same as the process of organization the creator of the work consciously experienced. Without an act of re-creation the object is not perceived as a work of art. [Dewey 1934a, p. 54; 1960, pp. 171–172]

And, we must add, without re-creation a proof is not experienced as a proof. Hermann Weyl, for one, insists emphatically on the distinction between a proof as an artifact, a lifeless "concatenation of grounds," and the insightful engagement with the artifact that yields the "experience of truth" [Weyl 1918, p. 11; 1994, p. 119]. Indeed, it is not enough "to get convicted, as it were, rather than convinced of a mathematical truth by a long chain of formal inferences and calculations leading us blindfolded from link to link" [Weyl 1985, p. 14]. There is a difference between confirming that a result follows and seeing why it follows.[15] A mathematical proof is not just an instrument of persuasion. It is the guise in which a mathematical experience haunts the world hoping to live anew in receptive minds. In Dewey's model at least, this new life requires a re-creation of creative experience. Re-creation may not measure up to creation but, then again, it ain't bad! I remember vividly the moment I was able to hold in my mind, all at once, a proof that every finitary closure space has a minimal closed basis. It would have been even more glorious if it had been *my* proof. But it was glorious nonetheless. There is a drive to *own* an insight, to grasp a solution in a grasping way. Puzzle drive, though, asks only that we come to understand, to survey the inner mechanism, to get at the *why*. It is wonderful just to *see* even when someone else shows us where to look and even when, to return to an earlier point, we have no idea how our insight might land us a job or cure a disease. Russell tells the story of Hobbes's first encounter with Euclid's geometry. As he worked through page after page, Hobbes was re-creating mathematics, not creating it. Nonetheless, says Russell, this was for Hobbes "a voluptuous moment," all the more so because "unsullied by the thought of the utility of geometry in measuring fields" [Russell 1932b, p. 31]. In sum, creative mathematical experiences are not only intrinsic goods; they are goods that live again in re-creations that are themselves intrinsically good. Re-creation is one form of propagation. There is another.

GROWTH

If we think of education as "a continuous process of reconstruction of experience," we view "every present experience as a moving force in influencing what future experiences will be" [Dewey 1963, p. 87]. An educative experience "arouses curiosity, strengthens initiative, and sets up desires and purposes that are sufficiently intense to carry a person over dead places in the future" [Dewey 1963, p. 38]. It can also supply skills, ideas, and habits of mind that help us turn problems into solutions. If, alongside problem-activity-success, there is also growth, then we emerge from a mathematical experience eager to pit heightened powers against even harder nuts. Yehuda Rav, for one, has already covered this ground both thoroughly and well. "Proofs are the mathematician's way to *display the mathematical machinery* for solving problems" [Rav 1999, p. 13]. It is only natural that mathematicians should offer such a display because "the essence of mathematics resides in inventing methods, tools, strategies and concepts for *solving problems*" [Rav 1999, p. 6]. To say that mathematicians are in the problem-solving business is, from Dewey's distinctive perspective, to say that they are in the business of educative experience—and that makes mathematics an affair of the deepest ethical significance. A coherent progression of ever richer experiences drawing on an expanding store of developing capacities: that is *the* good for a human being.

> There is an old saying to the effect that it is not enough for a man to be good; he must be good for something. The something for which a man must be good is capacity to live as a social member so that what he gets from living with others balances with what

he contributes. What he gets and gives as a human being, a being with desires, emotions, and ideas, is not external possessions, but a widening and deepening of conscious life—a more intense, disciplined, and expanding realization of meanings. What he materially receives and gives is at most opportunities and means for the evolution of conscious life. Otherwise, it is neither giving nor taking, but a shifting about of the position of things in space, like the stirring of water and sand with a stick. Discipline, culture, social efficiency, personal refinement, improvement of character are but phases of the growth of capacity nobly to share in such a balanced experience. And education is not a mere means to such a life. Education is such a life. To maintain capacity for such education is the essence of morals. [Dewey 1916, p. 417]

The socially valuable something for which a mathematician is good is, precisely, the dissemination of tools for the evolution of conscious life. Mathematicians have educative experiences: experiences that expand the capacity for educative experience. Their records of those experiences allow others to re-create them. Re-creation also brings growth: new purposes to challenge new capacities. Mathematicians grow in their capacity to grow and help others grow in their capacity to grow; those others can help others grow in their capacity to grow; and so on. Little wonder that mathematics displays "indefinite fertility" [Dewey 1938, p. 411]. Mathematical experiences, already treasures in themselves, propagate both by cloning (in re-creation) and by generating problems, techniques, and incentives that yield new educative experiences. If

$$\text{intrinsic value} + \text{contagious growth} = \text{goodness}$$

then our nonutilitarian panegyric of mathematics is complete. This equation is, however, wrong.

BREADTH

A safe presents the same challenge as a nut: it *must* be cracked. As a challenge to both hand and mind, it renews ancient demands on ancient resources. But *this* nut also offers problems intricate enough to hook a Nobel prize winner. Ignoring the economic side for the moment, the payoff can be a rich experience that strengthens our powers and gives zest to life. If you are a locksmith, you will share your knowledge with your apprentice. A criminal safecracker *might* do the same (like Peter Wimsey's friend Bill Rumm). Then growth leads to growth through straightforward transmission. Growth leads to growth in another way too: through an evolutionary arms race between safecracker and safe-maker. So safecracking, even in its illicit form, can display both intrinsic value and contagious growth. But, as Dewey is quick to suggest, a good safecracker will probably not be good. What is likely to be missing is growth that builds a cohesive structure of balanced experiences.

> That a man may grow in efficiency as a burglar, as a gangster, or as a corrupt politician, cannot be doubted. But from the standpoint of growth as education and education as growth the question is whether growth in this direction promotes or retards growth in general. Does this form of growth create conditions for further growth, or does it set up conditions that shut off the person who has grown in this particular direction from the occasions, stimuli, and opportunities for continuing growth in new directions? What is the effect of growth in a special direction upon the attitudes and habits which alone open up avenues for development in other lines? [Dewey 1963, p. 36]

Dewey does not elaborate on the likely trajectory of burglars, gangsters, and corrupt politicians, but he seems confident that the utmost perfection of their art only retards capacities whose exercise makes human life worth living. Instead of following the career of our safecracker, we return to our main interest: the mathematician. And here we confront an uncomfortable fact: mathematics as a practice and profession rewards the obsessive pursuit of results. The more obsessive this pursuit, the more other departments of life suffer. It does not follow, of course, that mathematics makes people obsessive. It is at least as likely that obsessives are drawn to mathematics because it offers a form of sublimation with tangible personal and social benefits. Furthermore, every occupation can be overdone. The hours you spend comforting dying orphans are less admirable if your own children are fending for themselves. Your prowess as a brain surgeon comes at a steep price if your single-minded devotion leaves you unable to enjoy Mozart. Hammers are good tools because they have a productive use, not because it is impossible to misuse them. Similarly, a line of work offers personal benefits if it *can* contribute an important piece to a rich and balanced human life. If we require that it cannot do otherwise, then no line of work counts as personally worthwhile. Mathematics may have a reputation for harboring especially odd characters because, in this field, glaring social inadequacies are compatible with professional success. A captain of industry, no matter how vicious in other respects, necessarily has a highly developed aptitude for successful social interaction in a variety of public settings. A mathematician who can manage only the poorest showing in public can still land a spot at the Institute for Advanced Study. An optimistic conclusion is that mathematics does not stunt variegated growth more than other vocations; it is just more tolerant of inadequacies that are glaringly obvious. In any case, it is clearly possible for broadly cultivated individuals to enjoy profound mathematical experiences. (Hermann Weyl and Bertrand Russell leap to mind.) So one essential feature of a balanced life, breadth, is compatible with intense devotion to mathematics. But balance is more than breadth. Balance also requires the right relationship between the various pieces of an expansive life experience.

INTEGRATION

In *Still Life*, Noël Coward tracks an adulterous affair from innocent start to dismal finish. Each of the play's five scenes finds the lovers, Alec and Laura, in the refreshment room of the Milford Junction train station: a place for breaks, interruptions; somewhere to pause on the way somewhere else; a place where no one lives. We soon perceive that the protagonists' situation is tragic because the pieces do not fit. First of all, the pieces of their shared experience do not fit *with one another* to form something larger; there is no hope that the episodes of the affair will add up to an integral experience. The only available consummations are physical ones leading nowhere, offering no growth, opening no new prospects. Alec and Laura keep arriving nowhere going nowhere. There is also no hope that the pieces of the affair will fit with the other pieces of their lives. "All the circumstances of our lives," says Alec, "have got to go on unaltered." So their love has to be "enclosed," walled off from everything else, "clean and untouched" [Coward 1935, p. 30]. Everyday values apply with full force everywhere else, but, to admit them into this enclave would be like using a yardstick to take a person's weight.

"This is different—something lovely and strange and desperately difficult. We can't measure it along with the values of our ordinary lives" [Coward 1935, p. 25]. The effort to maintain their love as a pause, a suspension, a static nowhere proves unsustainable both practically and psychologically. They cannot build the enclave walls high enough: the world intrudes. And Laura finds the dis-integration in which they seek protection reproducing itself alarmingly inside her own personality. "I love them just the same, Fred I mean and the children, but it's as though it wasn't me at all—as though I were looking on at someone else" [Coward 1935, p. 25]. In the end, it just ends. I mention all this because of its relevance to the doctrine of mathematical platonism: the view that mathematical objects are causally inert, supernatural beings with no coordinates in space or time. Whatever one may think of this as metaphysics, one must acknowledge a certain moral insight: mathematics provides a static nowhere, a refreshment room of the mind where, as long as the cat is fed and the children tucked in, we can enjoy voluptuous experiences without harm to any living creature. Causal inertness is a figure for a kind of ethical inertness. The point is emphatically not that mathematical experiences lack ethical significance. The point is rather that mathematical experiences have so little capacity to interact toxically with other valuable experiences. What Alec and Laura find so desperately difficult, mathematicians manage without effort, not because they have special powers but because no special powers are required. The ethical inertness of mathematical experiences only heightens their ethical significance. They add their special grace without any threat to the other blessings of a richly balanced life.

To return, finally, to our original question: is heightened control over physical and social forces really the only substantial contribution of mathematics to human prosperity? If mathematicians were soulless logic machines or if human happiness consisted in satisfaction of animal needs, this might have been plausible. If we accept, on the other hand, that our prosperity depends vitally on the character and consequences of our experiences and that mathematics offers intrinsically valuable experiences that can grow contagiously without disturbing the balance of our lives, then the above thesis becomes wholly implausible. The instrumental contribution of mathematics to our success is immense, but that is not the only way mathematics helps us flourish. All living beings harness energies and consume earthly goods. Our unrivaled

efficiency and rapacity mark an important quantitative distinction between ourselves and other species. More positively, we can be proud of
our success in the battle against "chaos without." We should, however,
be at least as proud of our victories over the "darkness within" in which
mathematics plays no small role. Mathematics pervades scientific explanations and, as we have seen, is also an engine of understanding, a
propagator of insightful experiences, in its own right.

This, too, is how we flourish, how we prosper, how we succeed as
individuals and as a species.

When Is Good Mathematics Good?

Though my panegyric of mathematics is now complete, I would like to
discuss one further issue: the question of whether nonmathematicians
have any legitimate role in the evaluation of mathematics. It is perfectly
clear that the answer to this question is "yes," and it will not take long
to show why.

Let us assume that there is a social consensus about who the mathematicians are. Then it seems convenient to let *them* decide what counts
as good mathematics and, indeed, what counts as mathematics at all.
We would be using a social consensus about the use of one term ("mathematician") to regiment our use of two related terms ("good mathematics" and "mathematics").

Now, however, we notice that good mathematics can be good *for*
various things—and it is not always an entirely mathematical question
whether a bit of good mathematics (a bit of mathematics deemed good
by the mathematicians) is good for some particular thing. Differential
geometers need help from structural geologists to determine whether
differential geometry has useful applications in structural geology. It is
not an entirely mathematical question whether that bit of good mathematics is good for that purpose. In general, the question of the instrumental utility of mathematics is not an entirely mathematical question.
So nonmathematicians have a legitimate role in the evaluation of mathematics—not, perhaps, *as* mathematics, but as a commodity that might
benefit humanity in various ways.

But this is not the end of the story: we have seen that instrumental utility is not the only contribution mathematics makes to human
happiness. Mathematics has additional ethical significance because it

makes further contributions to human prosperity. Our assessment of those contributions is highly sensitive to our understanding of ultimate values, our account of what it means for human beings to prosper. It would be perverse in the extreme to leave it to the mathematicians to provide that account. Our best articulations of ultimate values are the fruits of an ages-old conversation between poets and playwrights and priests and all sorts of people—even philosophers. This paper is an example of how philosophers might legitimately appraise mathematics—not *as* mathematics—but as an ally in the battle against Old Night.

There is no guarantee that every bit of good mathematics will hold up well under this scrutiny. It is conceivable that some good mathematics will be found to add little fuel to the lamp of reason. Hermann Weyl reached just this conclusion about classical set theory using a standard of undoubted philosophical parentage. He thought it a hallmark of intellectual responsibility to insist on an especially strong grasp of *Sinn* (meaning) as a precondition for an especially high degree of *Evidenz* (evidentness, the experience that a proposition expresses an evident truth).[16] If, for whatever reason, we reject Weyl's harsh assessment of classical set theory, we might yet agree that constructivist or intuitionist alternatives add more or different fuel to the lamp of reason. That, too, would be a philosophically grounded evaluation of some good mathematics.

Suppose we are presented with such an evaluation. How do we decide whether that evaluation is plausible? Well, we would have to *look*. We would have to subject the evaluation to scrutiny informed by philosophy, mathematics, and any number of things. It would not be intellectually responsible to insist, as a matter of principle, that no extramathematical appraisal of mathematics merits our attention.[17] Furthermore, friends of mathematics should not welcome such a grant of immunity since it allows only a desiccated account of how mathematics contributes to the good life. That is no great favor. Though some bits may fare worse than others, mathematics on the whole does very well, thank you, when measured against a full-blooded conception of human prosperity.

Notes

1. *Cf.* Yehuda Rav's effort "to stir the slumber of those who still think of mathematicians as deduction machines rather than creators of beautiful theories and inventors of methods and concepts to solve humanly meaningful problems" [Rav 1999, p. 17].

2. This flight of rhetoric will lead many of today's readers to wish that the language of the last century had soared in a more gender-neutral way.

3. See Graber [1995] for a thorough development of this point.

4. See also [1926, pp. 243–244], where Russell cites pure mathematics as a source of knowledge "valuable on its own account, quite apart from any use to which it is capable of being put." See Russell [1932b, ch. 2] for a sustained and positive appraisal of "useless" knowledge.

5. For Dewey's version of the same argument, see Dewey [1916, p. 283].

6. See also Russell [1932a, pp. 10–11] for Russell's inspired account of how Newton differs from an oyster: an account that will leave many in sympathy with the view that Newton, by "doing what is distinctively human," also "adds most to the diversified spectacle of nature."

7. ". . . to say that there are no such things as moral facts because desires control formation and valuation of ends is in truth but to point to desires and interests as themselves moral facts requiring control by intelligence equipped with knowledge" [Dewey 1939b, p. 154]. As Richard Bernstein puts it [1966, p. 125]: "Our values, our desires, our habits, can be nonrational or even irrational, but they are not forever cut off from our intelligence. They can be informed and transformed by intelligent deliberations." *Cf.* Dewey [1934b, pp. 48–50, 73]. See also Bernstein's discussion on pp. xxxii–xxxvi of Dewey [1960]. Dewey himself insists that philosophy has a role in the intelligent assessment of values because "Philosophy is criticism; criticism of the influential beliefs that underlie culture" [Dewey 1960, p. 107]. In this capacity, philosophy has "loosened the hold upon us exerted by predispositions that owe their strength to conformities which became so habitual as not to be questioned, and which in all probability would still be unquestioned were it not for the debt we owe to philosophers" [Dewey and Bentley 1949, p. 329; Dewey 1960, p. 149].

8. ". . . ideas are effective not as bare ideas but as they have imaginative content and emotional appeal. . . . The problem is that of effecting the union of ideas and knowledge with the non-rational factors in the human make-up. Art is the name given to all the agencies by which this union is effected" [Dewey 1939b, p. 150]. Art promotes "passionate intelligence": a union of "the heat of emotion and the light of intelligence" [Dewey 1934b, pp. 51–52, 79]. See also Dewey [1960, pp. 242–243].

9. Philosophers can make other contributions: for example, the "contagious diffusion of the scientific attitude" [Dewey 1939b, p. 153]. Such diffusion is *morally* significant because it makes our *moral* deliberations more intelligent. *Cf.* Bernstein [1966, pp. 122–125]. Note, too, Dewey's pursuit of "a logic, that is, a method of effective inquiry, which would apply without abrupt breach of continuity" to both science and morals [Dewey 1960, p. 15]. Another task for philosophy is to determine "the bearing of the conclusions reached in science . . . upon the value-factors involved in human action" [Dewey 1940b, p. 252; 1960, p. 254].

10. There is, by the way, material to ignore. Dewey devotes an entire chapter of [Dewey 1938] to "Mathematical discourse." See also Dewey [1929b, ch. 6] and Dewey [1960, p. 91]. See Dewey and Bentley [1949] for a definition of mathematics (p. 297) and a mature statement of the role of mathematics in human inquiry (pp. 318–321). For a clear and sympathetic summary of Dewey's philosophy of mathematics, see Downes [1961]. For an even-handed treatment of Dewey's approach to logic, see Nagel [1940]. For Russell's struggle to understand Dewey's notion of logic as theory of inquiry, see Russell [1939]. For Dewey's response, see Dewey [1939a].

11. If philosophy of art and philosophy of mathematics seem strange bedfellows, consider the following: Dewey's reflections on art seem to have raised his opinion of the sort of mental labor characteristic of mathematics. In *Democracy and Education* [1916], he declares that "fruitful understanding . . . cannot be attained purely mentally—just inside the head" (p. 321). This statement *might* express a view that Dewey still held when he wrote *Art as Experience* [1934a], but such a harsh formulation seems out of step with the general tendencies of that book—particularly Dewey's insistence on the intellectual gains available to someone whose experience of, say, a poem involves no agitation of body parts. Granted, some prior agitation probably set the stage for the experience. But if this movement is enough to disqualify an experience as purely mental, then one suspects that purely mental experiences are not only unfruitful but nonexistent.

12. For a helpful overview of Dewey's account of experience, see Bernstein [1966, chs. 4–7].

13. "Democracy is the faith that the process of experience is more important than any special result attained, so that special results achieved are of ultimate value only as they are used to enrich and order the ongoing process . . . Democracy as compared with other ways of life is the sole way of living which believes wholeheartedly in the process of experience as end and as means" [Dewey 1940a, pp. 227–228].

14. Note, too, Dewey's favorable assessment of Maurice Maeterlinck's idea that "experience has no goal save itself" [Dewey 1929a, p. 44].

15. See Dawson [2006] for a thorough elaboration of this point.

16. For discussion and references, see Pollard [2005].

17. Readers who suspect that I am tilting at windmills here might note Penelope Maddy's formulation of "naturalism" [Maddy 1997], particularly passages such as the following (p. 184): ". . . a successful enterprise, be it science or mathematics, should be understood and evaluated on its own terms . . . such an enterprise should not be subject to criticism from . . . some external, supposedly higher point of view . . . mathematics is not answerable to any extra-mathematical tribunal." If this just means that a bit of mathematics evaluated *as mathematics* should be evaluated on its own terms (that is, *as mathematics*), one can hardly quarrel with it. But then the blanket rejection of "any extra-mathematical tribunal" seems overstated. For another critical assessment of this version of naturalism, see Dieterle [1999]. For a helpful overview of Maddy's more recent position, see Shapiro and Reeder [2009].

References

Bernstein, Richard J. [1966]: *John Dewey*. New York: Washington Square Press.

Coward, Noël [1935]: *Still Life*. New York: Samuel French.

Dawson, John W. [2006]: "Why do mathematicians re-prove theorems?" *Philosophia Mathematica* (3)**14**, 269–86.

Dewey, John [1916]: *Democracy and Education*. New York: Macmillan.

Dewey, John [1929a]: *Characters and Events*, Vol. 1. New York: Henry Holt and Company.

Dewey, John [1929b]: *The Quest for Certainty*. New York: Minton, Balch and Company.

Dewey, John [1931]: *Philosophy and Civilization*. New York: Minton, Balch and Company.

Dewey, John [1934a]: *Art as Experience*. New York: Minton, Balch and Company.

Dewey, John [1934b]: *A Common Faith*. New Haven, CT: Yale University Press.

Dewey, John [1938]: *Logic: The Theory of Inquiry*. New York: Holt, Rinehart and Winston.

Dewey, John [1939a]: "Experience, knowledge and value: A rejoinder," in Schilpp [1939], pp. 515–608.

Dewey, John [1939b]: *Freedom and Culture*. New York: Putnam.

Dewey, John [1940a]: "Creative democracy—The task before us," in Ratner [1940], p. 227.

Dewey, John [1940b]: "Nature in experience," *Philosophical Review*, **49**, 244–58.

Dewey, John [1960]: *On Experience, Nature, and Freedom*. New York: Bobbs-Merrill.

Dewey, John [1963]: *Experience and Education*. New York: Collier Books.

Dewey, John, and Arthur F. Bentley [1949]: *Knowing and the Known*. Boston: Beacon Press.

Dieterle, J. M. [1999]: "Mathematical, astrological, and theological naturalism," *Philosophia Mathematica* (2)**7**, 129–35.

Downes, Chauncey B. [1961]: "Some problems concerning Dewey's view of reason," *Journal of Philosophy* **58**, 121–37.

Graber, Robert Bates [1995]: *Valuing Useless Knowledge*. Kirksville, MO: Thomas Jefferson University Press.

Maddy, Penelope [1997]: *Naturalism in Mathematics*. Oxford, U.K.: Clarendon Press.

Nagel, Ernest [1940]: "Dewey's reconstruction of logical theory," in Ratner [1940], pp. 56–86.

Pollard, Stephen [2005]: "Property is prior to set: Fichte and Weyl," in G. Sica, ed., *Essays on the Foundations of Mathematics and Logic*, Vol. 1, pp. 209–26. Monza, Italy: Polimetrica.

Pólya, G. [1957]: *How to solve it*. Garden City, NY: Doubleday Anchor Books.

Ratner, Sidney, ed. [1940]: *The Philosopher of the Common Man: Essays in Honor of John Dewey to Celebrate His Eightieth Birthday*. New York: Putnam.

Rav, Yehuda [1999]: "Why do we prove theorems?" *Philosophia Mathematica* (1)**7**, 5–41.

Russell, Bertrand [1926]: *Education and the Good Life*. New York: Boni and Liveright.

Russell, Bertrand [1932a]: *Education and the Social Order*. London: Allen and Unwin.

Russell, Bertrand [1932b]: *In Praise of Idleness*. London: Allen and Unwin.

Russell, Bertrand [1935]: *Religion and Science*. London: Thornton Butterworth.

Russell, Bertrand [1939]: "Dewey's new *Logic*," in Schilpp [1939], pp. 135–56.

Schilpp, P. A., ed. [1939]: *The Philosophy of John Dewey*. Evanston, IL: Northwestern University Press.

Shapiro, Stewart, and Patrick Reeder [2009]: "A scientific enterprise?: Penelope Maddy's *Second Philosophy*," *Philosophia Mathematica* (2)**17**, 247–71.

Weyl, Hermann [1918]: *Das Kontinuum*. Leipzig, Germany: Veit.

Weyl, Hermann [1985]: "Axiomatic versus constructive procedures in mathematics," *Mathematical Intelligencer* **7**(4), 12–17, 38.

Weyl, Hermann [1994]: *The Continuum*. New York: Dover.

Credit: "Mathematics and the Good Life" by Stephen Pollard. *Philosophia Mathematica* 21.1(2013): 93–109, by permission of Oxford University Press.

The Rise of Big Data: How It's Changing the Way We Think about the World

Kenneth Cukier and Viktor Mayer-Schönberger

Everyone knows that the Internet has changed how businesses operate, governments function, and people live. But a new, less visible technological trend is just as transformative: "big data." Big data starts with the fact that there is a lot more information floating around these days than ever before, and it is being put to extraordinary new uses. Big data is distinct from the Internet, although the web makes it much easier to collect and share data. Big data is about more than just communication: the idea is that we can learn from a large body of information things that we could not comprehend when we used only smaller amounts.

In the third century BC, the Library of Alexandria was believed to house the sum of human knowledge. Today, there is enough information in the world to give every person alive 320 times as much of it as historians think was stored in Alexandria's entire collection—an estimated 1,200 exabytes' worth. If all this information were placed on CDs and they were stacked up, the CDs would form five separate piles that would all reach to the moon.

This explosion of data is relatively new. As recently as the year 2000, only one-quarter of all the world's stored information was digital. The rest was preserved on paper, film, and other analog media. But because the amount of digital data expands so quickly—doubling around every three years—that situation was swiftly inverted. Today, less than two percent of all stored information is nondigital.

Given this massive scale, it is tempting to understand big data solely in terms of size. But that would be misleading. Big data is also characterized by the ability to render into data many aspects of the world that have never been quantified before; call it "datafication." For example, location has been datafied, first with the invention of longitude and latitude, and more recently with GPS satellite systems. Words are treated as data when computers mine centuries' worth of books. Even friendships and "likes" are datafied, via Facebook.

This kind of data is being put to incredible new uses with the assistance of inexpensive computer memory, powerful processors, smart algorithms, clever software, and math that borrows from basic statistics. Instead of trying to "teach" a computer how to do things, such as drive a car or translate between languages, which artificial intelligence experts have tried unsuccessfully to do for decades, the new approach is to feed enough data into a computer that it can infer the probability that, say, a traffic light is green and not red or that, in a certain context, *lumière* is a more appropriate substitute for "light" than *léger*.

Using great volumes of information in this way requires three profound changes in how we approach data. The first is to collect and use a lot of data rather than settle for small amounts or samples, as statisticians have done for well over a century. The second is to shed our preference for highly curated and pristine data and instead accept messiness: in an increasing number of situations, a bit of inaccuracy can be tolerated because the benefits of using vastly more data of variable quality outweigh the costs of using smaller amounts of very exact data. Third, in many instances, we need to give up our quest to discover the cause of things, in return for accepting correlations. With big data, instead of trying to understand precisely why an engine breaks down or why a drug's side effect disappears, researchers can instead collect and analyze massive quantities of information about such events and everything that is associated with them, looking for patterns that might help predict future occurrences. Big data helps answer what, not why, and often that's good enough.

The Internet has reshaped how humanity communicates. Big data is different: it marks a transformation in how society processes information. In time, big data might change our way of thinking about the world. As we tap ever more data to understand events and make decisions, we are likely to discover that many aspects of life are probabilistic, rather than certain.

Approaching "n = all"

For most of history, people have worked with relatively small amounts of data because the tools for collecting, organizing, storing, and analyzing information were poor. People winnowed the information they relied on to the barest minimum so that they could examine it more easily. This was the genius of modern-day statistics, which first came to the fore in the late nineteenth century and enabled society to understand complex realities even when few data existed. Today, the technical environment has shifted 179 degrees. There still is, and always will be, a constraint on how much data we can manage, but it is far less limiting than it used to be and will become even less so as time goes on.

The way people handled the problem of capturing information in the past was through sampling. When collecting data was costly and processing it was difficult and time-consuming, the sample was a savior. Modern sampling is based on the idea that, within a certain margin of error, one can infer something about the total population from a small subset, as long the sample is chosen at random. Hence, exit polls on election night query a randomly selected group of several hundred people to predict the voting behavior of an entire state. For straightforward questions, this process works well. But it falls apart when we want to drill down into subgroups within the sample. What if a pollster wants to know which candidate single women under 30 are most likely to vote for? How about university-educated, single Asian American women under 30? Suddenly, the random sample is largely useless, since there may be only a couple of people with those characteristics in the sample, too few to make a meaningful assessment of how the entire subpopulation will vote. But if we collect all the data—"n = all," to use the terminology of statistics—the problem disappears.

This example raises another shortcoming of using some data rather than all of it. In the past, when people collected only a few data, they often had to decide at the outset what to collect and how it would be used. Today, when we gather all the data, we do not need to know beforehand what we plan to use it for. Of course, it might not always be possible to collect all the data, but it is getting much more feasible to capture vastly more of a phenomenon than simply a sample and to aim for all of it. Big data is a matter not just of creating somewhat larger samples but of harnessing as much of the existing data as possible about what is being studied. We still need statistics; we just no longer need to rely on small samples.

There is a trade-off to make, however. When we increase the scale by orders of magnitude, we might have to give up on clean, carefully curated data and tolerate some messiness. This idea runs counter to how people have tried to work with data for centuries. Yet the obsession with accuracy and precision is in some ways an artifact of an informationconstrained environment. When there were not that many data around, researchers had to make sure that the figures they bothered to collect were as exact as possible. Tapping vastly more data means that we can now allow some inaccuracies to slip in (provided the data set is not completely incorrect), in return for benefiting from the insights that a massive body of data provides.

Consider language translation. It might seem obvious that computers would translate well, since they can store lots of information and retrieve it quickly. But if one were to simply substitute words from a French-English dictionary, the translation would be atrocious. Language is complex. A breakthrough came in the 1990s, when IBM delved into statistical machine translation. It fed Canadian parliamentary transcripts in both French and English into a computer and programmed it to infer which word in one language is the best alternative for another. This process changed the task of translation into a giant problem of probability and math. But after this initial improvement, progress stalled.

Then Google barged in. Instead of using a relatively small number of high-quality translations, the search giant harnessed more data, but from the less orderly Internet—"data in the wild," so to speak. Google inhaled translations from corporate websites, documents in every language from the European Union, even translations from its giant bookscanning project. Instead of millions of pages of texts, Google analyzed billions. The result is that its translations are quite good—better than IBM's were—and cover 65 languages. Large amounts of messy data trumped small amounts of cleaner data.

From Causation to Correlation

These two shifts in how we think about data—from some to all and from clean to messy—give rise to a third change: from causation to correlation. This change represents a move away from always trying to understand the deeper reasons behind how the world works to simply learning about an association among phenomena and using that to get things done.

Of course, knowing the causes behind things is desirable. The problem is that causes are often extremely hard to figure out, and many times, when we think we have identified them, it is nothing more than a self-congratulatory illusion. Behavioral economics has shown that humans are conditioned to see causes even where none exist. So we need to be particularly on guard to prevent our cognitive biases from deluding us; sometimes, we just have to let the data speak.

Take UPS, the delivery company. It places sensors on vehicle parts to identify certain heat or vibrational patterns that in the past have been associated with failures in those parts. In this way, the company can predict a breakdown before it happens and replace the part when it is convenient, instead of on the side of the road. The data do not reveal the exact relationship between the heat or the vibrational patterns and the part's failure. They do not tell UPS why the part is in trouble. But they reveal enough for the company to know what to do in the near term and guide its investigation into any underlying problem that might exist with the part in question or with the vehicle.

A similar approach is being used to treat breakdowns of the human machine. Researchers in Canada are developing a big-data approach to spot infections in premature babies before overt symptoms appear. By converting 16 vital signs, including heartbeat, blood pressure, respiration, and blood-oxygen levels, into an information flow of more than 1,000 data points per second, they have been able to find correlations between very minor changes and more serious problems. Eventually, this technique will enable doctors to act earlier to save lives. Over time, recording these observations might also allow doctors to understand what actually causes such problems. But when a newborn's health is at risk, simply knowing that something is likely to occur can be far more important than understanding exactly why.

Medicine provides another good example of why, with big data, seeing correlations can be enormously valuable, even when the underlying causes remain obscure. In February 2009, Google created a stir in healthcare circles. Researchers at the company published a paper in *Nature* that showed how it was possible to track outbreaks of the seasonal flu using nothing more than the archived records of Google searches. Google handles more than a billion searches in the United States every day and stores them all. The company took the 50 million most commonly searched terms between 2003 and 2008 and compared them against historical influenza data from the Centers for Disease Control and Prevention. The idea was to discover whether the incidence of certain searches coincided with outbreaks of the flu—in other words, to see whether an increase in the frequency of certain Google searches conducted in a particular geographic area correlated with the CDC's data on outbreaks of flu there. The CDC tracks actual patient visits to hospitals and clinics across the country, but the information it releases suffers from a reporting lag of a week or two—an eternity in the case of a pandemic. Google's system, by contrast, would work in near-real time.

Google did not presume to know which queries would prove to be the best indicators. Instead, it ran all the terms through an algorithm that ranked how well they correlated with flu outbreaks. Then the system tried combining the terms to see if that improved the model. Finally, after running nearly half a billion calculations against the data, Google identified 45 terms—words such as "headache" and "runny nose"—that had a strong correlation with the CDC's data on flu outbreaks. All 45 terms related in some way to influenza. But with a billion searches a day, it would have been impossible for a person to guess which ones might work best and test only those.

Moreover, the data were imperfect. Since the data were never intended to be used in this way, misspellings and incomplete phrases were common. But the sheer size of the data set more than compensated for its messiness. The result, of course, was simply a correlation. It said nothing about the reasons why someone performed any particular search. Was it because the person felt ill, or heard sneezing in the next cubicle, or felt anxious after reading the news? Google's system doesn't know, and it doesn't care. Indeed, last December, it seems that Google's system may have overestimated the number of flu cases in the United States. This serves as a reminder that predictions are only probabilities and are not always correct, especially when the basis for the prediction—Internet searches—is in a constant state of change and vulnerable to outside influences, such as media reports. Still, big data can hint at the general direction of an ongoing development, and Google's system did just that.

Back-End Operations

Many technologists believe that big data traces its lineage back to the digital revolution of the 1980s, when advances in microprocessors and computer memory made it possible to analyze and store ever more information. That is only superficially the case. Computers and the Internet certainly aid big data by lowering the cost of collecting, storing, processing, and sharing information. But at its heart, big data is only the latest step in humanity's quest to understand and quantify the world. To appreciate how this is the case, it helps to take a quick look behind us.

Appreciating people's posteriors is the art and science of Shigeomi Koshimizu, a professor at the Advanced Institute of Industrial Technology in Tokyo. Few would think that the way a person sits constitutes information, but it can. When a person is seated, the contours of the body, its posture, and its weight distribution can all be quantified and tabulated. Koshimizu and his team of engineers convert backsides into data by measuring the pressure they exert at 360 different points with sensors placed in a car seat and by indexing each point on a scale of zero to 256. The result is a digital code that is unique to each individual. In a trial, the system was able to distinguish among a handful of people with 98 percent accuracy.

The research is not asinine. Koshimizu's plan is to adapt the technology as an antitheft system for cars. A vehicle equipped with it could recognize when someone other than an approved driver sat down behind the wheel and could demand a password to allow the car to function. Transforming sitting positions into data creates a viable service and a potentially lucrative business. And its usefulness may go far beyond deterring auto theft. For instance, the aggregated data might reveal clues about a relationship between drivers' posture and road safety, such as telltale shifts in position before accidents. The system might also be able to sense when a driver slumps slightly from fatigue and send an alert or automatically apply the brakes.

Koshimizu took something that had never been treated as data—or even imagined to have an informational quality—and transformed it into a numerically quantified format. There is no good term yet for this sort of transformation, but "datafication" seems apt. Datafication is not the same as digitization, which takes analog content—books, films, photographs—and converts it into digital information, a sequence of ones and zeros that computers can read. Datafication is a far broader activity: taking all aspects of life and turning them into data. Google's augmented-reality glasses "datafy" the gaze. Twitter datafies stray thoughts. LinkedIn datafies professional networks.

Once we datafy things, we can transform their purpose and turn the information into new forms of value. For example, IBM was granted a U.S. patent in 2012 for "securing premises using surfacebased computing technology"—a technical way of describing a touchsensitive floor covering, somewhat like a giant smartphone screen. Datafying the floor can open up all kinds of possibilities. The floor could be able to identify the objects on it, so that it might know to turn on lights in a room or open doors when a person entered. Moreover, it might identify individuals by their weight or by the way they stand and walk. It could tell if someone fell and did not get back up, an important feature for the elderly. Retailers could track the flow of customers through their stores. Once it becomes possible to turn activities of this kind into data that can be stored and analyzed, we can learn more about the world— things we could never know before because we could not measure them easily and cheaply.

Big Data in the Big Apple

Big data will have implications far beyond medicine and consumer goods: it will profoundly change how governments work and alter the nature of politics. When it comes to generating economic growth, providing public services, or fighting wars, those who can harness big data effectively will enjoy a significant edge over others. So far, the most exciting work is happening at the municipal level, where it is easier to access data and to experiment with the information. In an effort spearheaded by New York City Mayor Michael Bloomberg (who made a fortune in the data business), the city is using big data to improve public services and lower costs. One example is a new fire-prevention strategy.

Illegally subdivided buildings are far more likely than other buildings to go up in flames. The city gets 25,000 complaints about overcrowded buildings a year, but it has only 200 inspectors to respond. A small team of analytics specialists in the mayor's office reckoned that big data could help resolve this imbalance between needs and resources. The team created a database of all 900,000 buildings in the city and augmented it with troves of data collected by 19 city agencies: records of tax liens, anomalies in utility usage, service cuts, missed payments, ambulance visits, local crime rates, rodent complaints, and more. Then they compared this database to records of building fires from the past five years, ranked by severity, hoping to uncover correlations. Not surprisingly, among the predictors of a fire were the type of building and the year it was built. Less expected, however, was the finding that buildings that obtained permits for exterior brickwork correlated with lower risks of severe fire.

Using all this data allowed the team to create a system that could help them determine which overcrowding complaints needed urgent attention. None of the buildings' characteristics they recorded caused fires; rather, they correlated with an increased or decreased risk of fire. That knowledge has proved immensely valuable: in the past, building inspectors issued vacate orders in 13 percent of their visits; using the new method, that figure rose to 70 percent—a huge efficiency gain.

Of course, insurance companies have long used similar methods to estimate fire risks, but they mainly rely on only a handful of attributes and usually ones that intuitively correspond with fires. By contrast, New York City's big-data approach was able to examine many more variables, including ones that would not at first seem to have any relation to fire risk. And the city's model was cheaper and faster, since it made use of existing data. Most important, the big-data predictions are probably more on target, too.

Big data is also helping increase the transparency of democratic governance. A movement has grown up around the idea of "open data," which goes beyond the freedom-of-information laws that are now commonplace in developed democracies. Supporters call on governments to make the vast amounts of innocuous data that they hold easily available to the public. The United States has been at the forefront, with its data.gov website, and many other countries have followed.

At the same time as governments promote the use of big data, they also need to protect citizens against unhealthy market dominance. Companies such as Google, Amazon, and Facebook—as well as lesser known "data brokers," such as Acxiom and Experian—are amassing vast amounts of information on everyone and everything. Antitrust laws protect against the monopolization of markets for goods and services such as software or media outlets because the sizes of the markets for those goods are relatively easy to estimate. But how should governments apply antitrust rules to big data, a market that is hard to define and that is constantly changing form? Meanwhile, privacy will become an even bigger worry, since more data will almost certainly lead to more compromised private information, a downside of big data that current technologies and laws seem unlikely to prevent.

Regulations governing big data might even emerge as a battleground among countries. European governments are already scrutinizing Google over a raft of antitrust and privacy concerns in a scenario reminiscent of the antitrust enforcement actions the European Commission took against Microsoft beginning a decade ago. Facebook might become a target for similar actions all over the world because it holds so much data about individuals. Diplomats should brace for fights over whether to treat information flows as similar to free trade: in the future, when China censors Internet searches, it might face complaints not only about unjustly muzzling speech but also about unfairly restraining commerce.

Big Data or Big Brother?

States will need to help protect their citizens and their markets from new vulnerabilities caused by big data. But there is another potential dark side: big data could become Big Brother. In all countries, but particularly in nondemocratic ones, big data exacerbates the existing asymmetry of power between the state and the people.

The asymmetry could well become so great that it leads to big-data authoritarianism, a possibility vividly imagined in science fiction movies such as *Minority Report*. That 2002 film took place in a nearfuture dystopia in which the character played by Tom Cruise headed a "Precrime" police unit that relied on clairvoyants whose visions identified people who were about to commit crimes. The plot revolves around the system's obvious potential for error and, worse yet, its denial of free will.

Although the idea of identifying potential wrongdoers before they have committed a crime seems fanciful, big data has allowed some authorities to take it seriously. In 2007, the Department of Homeland Security launched a research project called FAST (Future Attribute Screening Technology), aimed at identifying potential terrorists by analyzing data about individuals' vital signs, body language, and other physiological patterns. Police forces in many cities, including Los Angeles, Memphis, Richmond, and Santa Cruz, have adopted "predictive policing" software, which analyzes data on previous crimes to identify where and when the next ones might be committed.

For the moment, these systems do not identify specific individuals as suspects. But that is the direction in which things seem to be heading. Perhaps such systems would identify which young people are most likely to shoplift. There might be decent reasons to get so specific, especially when it comes to preventing negative social outcomes other than crime. For example, if social workers could tell with 95 percent accuracy which teenage girls would get pregnant or which high school boys would drop out of school, wouldn't they be remiss if they did not step in to help? It sounds tempting. Prevention is better than punishment, after all. But even an intervention that did not admonish and instead provided assistance could be construed as a penalty—at the very least, one might be stigmatized in the eyes of others. In this case, the state's actions would take the form of a penalty before any act were committed, obliterating the sanctity of free will.

Another worry is what could happen when governments put too much trust in the power of data. In his 1999 book, *Seeing Like a State,* the anthropologist James Scott documented the ways in which governments, in their zeal for quantification and data collection, sometimes end up making people's lives miserable. They use maps to determine how to reorganize communities without first learning anything about the people who live there. They use long tables of data about harvests to decide to collectivize agriculture without knowing a whit about farming. They take all the imperfect, organic ways in which people have interacted over time and bend them to their needs, sometimes just to satisfy a desire for quantifiable order.

This misplaced trust in data can come back to bite. Organizations can be beguiled by data's false charms and endow more meaning to the numbers than they deserve. That is one of the lessons of the Vietnam War. U.S. Secretary of Defense Robert McNamara became obsessed with using statistics as a way to measure the war's progress. He and his colleagues fixated on the number of enemy fighters killed. Relied on by commanders and published daily in newspapers, the body count became the data point that defined an era. To the war's supporters, it was proof of progress; to critics, it was evidence of the war's immorality. Yet the statistics revealed very little about the complex reality of the conflict. The figures were frequently inaccurate and were of little value as a way to measure success. Although it is important to learn from data to improve lives, common sense must be permitted to override the spreadsheets.

Human Touch

Big data is poised to reshape the way we live, work, and think. A world-view built on the importance of causation is being challenged by a preponderance of correlations. The possession of knowledge, which once meant an understanding of the past, is coming to mean an ability to predict the future. The challenges posed by big data will not be easy to resolve. Rather, they are simply the next step in the timeless debate over how to best understand the world.

Still, big data will become integral to addressing many of the world's pressing problems. Tackling climate change will require analyzing pollution data to understand where best to focus efforts and find ways to mitigate problems. The sensors being placed all over the world, including those embedded in smartphones, provide a wealth of data that will allow climatologists to more accurately model global warming. Meanwhile, improving and lowering the cost of health care, especially for the world's poor, will make it necessary to automate some tasks that currently require human judgment but could be done by a computer, such as examining biopsies for cancerous cells or detecting infections before symptoms fully emerge.

Ultimately, big data marks the moment when the "information society" finally fulfills the promise implied by its name. The data take center stage. All those digital bits that have been gathered can now be harnessed in novel ways to serve new purposes and unlock new forms of value. But this situation requires a new way of thinking and will challenge institutions and identities. In a world where data shape decisions more and more, what purpose will remain for people, or for intuition, or for going against the facts? If everyone appeals to the data and harnesses big-data tools, perhaps what will become the central point of differentiation is unpredictability: the human element of instinct, risk taking, accidents, and even error. If so, then there will be a special need to carve out a place for the human: to reserve space for intuition, common sense, and serendipity to ensure that they are not crowded out by data and machine-made answers.

This possibility has important implications for the notion of progress in society. Big data enables us to experiment faster and explore more leads. These advantages should produce more innovation. But at times, the spark of invention becomes what the data do not say. That is something that no amount of data can ever confirm or corroborate, since it has yet to exist. If Henry Ford had queried big-data algorithms to discover what his customers wanted, they would have come back with "a faster horse," to recast his famous line. In a world of big data, it is the most human traits that will need to be fostered—creativity, intuition, and intellectual ambition—since human ingenuity is the source of progress.

Big data is a resource and a tool. It is meant to inform, rather than explain; it points toward understanding, but it can still lead to misunderstanding, depending on how well it is wielded. And however dazzling the power of big data appears, its seductive glimmer must never blind us to its inherent imperfections. Rather, we must adopt this technology with an appreciation not just of its power but also of its limitations.

Conway's Wizards

Tanya Khovanova

This article is about a puzzle invented by John H. Conway. Conway e-mailed it to me in June 2009. I wrote about it on my blog [1]. I also invented and posted a simplified version [2] and a generalized version [3]. But I never published a solution or a discussion. Now the time has come to do so.

The Original Puzzle

John Conway sent me a puzzle about wizards, which he invented in the 1960s. Here it is:

Last night I sat behind two wizards on a bus and overheard the following:

A: "I have a positive integral number of children, whose ages are positive integers, the sum of which is the number of this bus, while the product is my own age."

B: "How interesting! Perhaps if you told me your age and the number of your children, I could work out their individual ages?"

A: "No."

B: "Aha! AT LAST I know how old you are!'"

Now what was the number of the bus?

A Discussion and a Solution

This is an incredible puzzle. This is also an underappreciated puzzle. It is more interesting than it might seem. When someone announces the answer, it is not clear whether they have solved it completely. No wonder Conway looked at me with suspicion when, the day after receiving the puzzle, I told him the number of the bus.

I am a spoiler. I will spoil you and will tell you how to solve this puzzle, but not right now. First, I want to apologize for the seeming rudeness of wizard A. When he says, "No," he does not mean, "I do not want to tell you my age and the number of my children." "No" is the answer to the previous question, and it means, "My age and the number of my children are not enough to imply their individual ages."

Here is my real question. Why were the wizards riding the bus? Could they have been on a trolley? No, they could not have been. Conway designs his problems very carefully. In this puzzle, he implicitly introduced a good notation. I will follow his suggestion and denote the age of the wizard, the number of the bus, and the number of the children by a, b, and c.

EXAMPLES

Let us start with some examples. Suppose $b = 5$. Here is the list of possible ages for children, and the corresponding age of the wizard and the number of children:

- 1, 1, 1, 1, and 1. Then $a = 1$ and $c = 5$.
- 1, 1, 1, and 2. Then $a = 2$ and $c = 4$.
- 1, 1, and 3. Then $a = 3$ and $c = 3$.
- 1, 2, and 2. Then $a = 4$ and $c = 3$.
- 1 and 4. Then $a = 4$ and $c = 2$.
- 5. Then $a = 5$ and $c = 1$.

For the purpose of the theoretical discussion, let us ignore the ridiculously small age of the wizard. We see that if the wizard is not 4 years old, his age is sufficient to determine the number and the ages of the children. In any case, the wizard's age and the number of his children uniquely determine their ages. So wizard A could not have said, "No." From this example, we can conclude that the bus number is not 5. Checking other small cases, we can assure ourselves that the bus number is greater than 5: $b > 5$.

We can keep increasing the bus number by 1 until we find the bus number for which the wizard A could have said "No." One might think this gives us the answer. We will obtain a right answer this way and produce a correct bus number and the age of the wizard, but we will miss a big part of the story.

Let us look at large bus numbers. Suppose that $b = 21$. The wizard A can say "No" in this case. Indeed, if he were 96 and had three children, their ages are not uniquely defined. They could be 1, 8, and 12, or alternatively, 2, 3, and 16. But the important question is whether the wizard B can now determine that A is 96. Not really. It might be that the wizard A is 240 and has three children: 4, 5, and 12, or alternatively, 3, 8, and 10. So 21 cannot be the bus number, but for a completely different reason. Wizard B could not have figured out the age of wizard A.

Does this always happen with large numbers? Let us try 22. If we think about it, we do not even need to perform new calculations. The wizard's age is ambiguous again. He can be 96 and can have four children of ages 1, 1, 8, and 12, or 1, 2, 3, and 16. Similarly, he can be 240 and have four children 1, 4, 5, and 12, or alternatively, 1, 3, 8, and 10. Do you see what happened? We can add one more child of age 1. This way, the number of children increases by 1, the sum increases by 1, and the product stays the same. That means that if bus b had two possible ages for wizard B, then the same two ages would work for bus $b + 1$. Therefore, we do not need to check large numbers that are greater than 21: $b < 21$.

Now our problem becomes finite. We just need to try all the possibilities up to 21 to find the answer. It is lucky for Conway that the bus number is uniquely defined in this way. It could have been that as soon as wizard A said "No," the age would not be uniquely defined. Or, it could have been that there were several bus numbers for which wizard B could have guessed A's age.

Now to the answer. The bus number is 12. The only age for which the bus number and the number of children do not define the ages of children is 48. The children's ages could be 2, 2, 2, and 6, or, on the other hand, 1, 3, 4, and 4.

To complete the solution, we need to check that all the bus numbers less than 12 will not permit wizard A to say no, and bus numbers greater than 12 will not define the age of wizard A uniquely. I will leave the first task to you. I want to show that bus numbers starting from 13 do not work. Indeed, if the bus number is 13, wizard A can say "no" if he is 48 and has five children, aged either 1, 2, 2, 2, and 6, or 1, 1, 3,

4, and 4. The second possible age for wizard A is 36, with three children aged either 2, 2, and 9, or 1, 6, and 6. If the bus number is 13 or greater, wizard B cannot figure out the age of wizard A.

THE NUMBER OF CHILDREN

Here I would like to reflect on the number of children.

It is obvious from the start that wizard A has more than two children. If he had one child, then the child's age would be both the number of the bus and the same as the father's age. Although this is unrealistic, in mathematics strange things can happen. The important part is that if wizard A has one child he could not have said "No." The same is true for two children: Their age distribution is uniquely defined by the sum and the product of their ages.

So wizard A has to have at least three children. One common mistake is to assume that wizard A has to have exactly three children. People who make this mistake might well discover the beauty of this puzzle, but they'll miss the bus, so to speak.

These people will find that the first case, when wizard A says "No" and has three children, is for bus 13, when wizard A's age is 36, and his children's ages are either 2, 2, and 9, or 1, 6, and·6. The next bus number, 14, produces two potential ages for wizard A. First, he could be 40 and his children could be aged either 2, 2, and 10, or 1, 5, and 8. Second, he could be 72 and his children could be aged either 3, 3, and 8, or 2, 6, and 6. So if someone tells you that the answer is 13, you can safely guess that that person assumed that wizard A has three children.

Simplified Wizards

Now that we have discussed the solution to Conway's puzzle, I would like to share a simpler puzzle that you will have to solve on your own. For the sake of continuity, I suppressed reality; assume that ages do not need to be realistic.

Last night I sat behind two wizards on a bus and overheard the following:

> A: "I have a positive integral number of children, whose ages are positive integers, the sum of which is the number of this bus, while the product is my own age."

B: "How interesting! Perhaps if you told me the number of your children, I could work out their individual ages?"

A: "No."

B: "Aha! AT LAST I know how many children you have!"

What is the number of the bus? While you are at it, is it possible to figure out the age of the wizard?

Surprisingly, the number of the bus can again be uniquely determined.

Generalized Wizards

Now I want to give you a more difficult puzzle:

Last night I sat behind two wizards on a bus and overheard the following:

A: "I have a positive integral number of children, whose ages are positive integers, the sum of which is the number of this bus, while the product is my own age. Also, the sum of the squares of their ages is the number of dolls in my collection."

B: "How interesting! Perhaps if you told me your age, the number of your children, and the number of dolls, I could work out your children's individual ages."

A: "No."

B: "Aha! AT LAST I know how old you are!"

Now, what was the number of the bus?

By the way, by now you should be able to figure out why I prefer "dolls" over "baseball cards."

Although I kept the focus of this puzzle on the bus and on the age of the wizard for the sake of continuity with Conway's original puzzle, I had to sacrifice realism. For this puzzle, you need to have an open mind. In Conway's original puzzle you do not need to assume that wizard A's age is in a particular range, but after you solve it, you see that his age makes sense. In this generalized puzzle, you might be surprised how long wizards can live and keep their fertility.

Another difference with the original puzzle is that it is difficult to solve this one without a computer.

Once again the solution is unique. The number of the bus is 26. All buses with numbers less than 25 do not give wizard A the opportunity to say "No." If the bus is 26, the wizard is 3,456 years old and has seven children and 124 dolls. The ages of the children are one of the following:

- 1, 3, 3, 3, 4, 4, 8
- 2, 2, 2, 2, 6, 6, 6

If the bus number is 27, we can reuse the solution for bus 26 and add one more child, aged 1. Thus wizard A could be 3,456 years of age, could have eight children, and could have 125 dolls. This bus number, however, has another solution for the younger age of 2560, six children, and 165 dolls:

- 1, 4, 4, 4, 4, 10
- 2, 2, 2, 5, 8, 8

You can try to continue to the next step of generalization and create another puzzle by adding the next symmetric polynomial on the ages of the children, for example, the sum of cubes. In this case, I do not know if the puzzle works: that is, if there is a unique bus number. I did not have the patience or the computer wisdom to do this. I conjecture that uniqueness will stop. Luck will run out, and we will not have a clean puzzle any longer.

References

[1] T. Khovanova, "John Conway's Wizards," available at http://blog.tanyakhovanova.com /?p=29 (2008).

[2] T. Khovanova, "Simplified Wizards Puzzle," available at http://blog. tanyakhovanova.com /?p=33 (2008).

[3] T. Khovanova, "Conway's Wizards Generalized," available at http://blog.tanyakhovanova .com/?p=36 (2008).

On Unsettleable Arithmetical Problems

John H. Conway

Before Fermat's Last Theorem was proved, there was some speculation that it might be unprovable. Many people noticed that the theorem and its negation have a different status. The negation asserts that for some $n > 2$ there is an nth power that is the sum of two smaller ones. Exhibiting these numbers proves the negation and disproves the theorem itself. So if one shows that the theorem is not disprovable, then one also shows that there exist no such nth powers and therefore that the theorem is true.

However, the theorem could conceivably be true without being provable. In this case, its unprovability could not itself be proved since such a proof would imply the nonexistence of a counterexample.

The same sort of arguments applied to the Four Color Theorem and still apply to Goldbach's Conjecture, which states that every even number greater than 2 is the sum of two primes. (In fact, Goldbach asserted this of every positive even number since he counted 1 as a prime.) There has never been any doubt that Goldbach's conjecture is true because the evidence for it is overwhelming.

What are the simplest true assertions that are neither provable nor disprovable? I shall use the term unsettleable because for more than a century the ultimate basis for proof has been set theory. For some of my examples, it might even be that the assertion that they are not provable is not itself provable, and so on. Of course, this means that you shouldn't expect to see any proofs! My examples are inspired by the Collatz $3n + 1$ problem.

The Collatz 3n + 1 Problem

Consider the Collatz function $(1/2)n \mid 3n + 1$, whose value is $(1/2)n$ if this is an integer and otherwise $3n + 1$. I shall call this a "bipartite linear function" because its value is one of two linear possibilities. The

Collatz $3n + 1$ problem is, "Does iterating this function always eventually lead to 1" (starting at a positive integer)? It certainly does if we start at 7:

$$7 \to 22 \to 11 \to 34 \to 17 \to 52 \to 26 \to 13 \to$$
$$40 \to 20 \to 10 \to 5 \to 16 \to 8 \to 4 \to 2 \to 1.$$

Tomás Oliveira e Silva has verified [4] that it does for all numbers less than 5×10^{18}. There is a slight chance that this problem itself is unsettleable—some similar problems certainly are.

I generalize it by considering multipartite linear functions and the associated games and problems. The value of the k-partite linear function

$$g(n) = g_1(n) \, |g_2(n)| \, \ldots \, | \, g_k(n)$$

is the first one of the k linear functions $g_i(n) = a_i n + b_i$ that is integral (and is undefined if no $g_i(n)$ is integral). The corresponding *Collatzian game* is to repeatedly replace n by $g(n)$ until a predetermined number (1, say) is reached, or possibly $g(n)$ is undefined, when the game stops.

Are There Unsettleable Collatzian Games?

There certainly are. The proof is more technical than the rest of the paper, but the message is simple: There is an explicit game with 24 simple linear functions for which there are numbers n for which the game never stops, but this is not provable. Gödel's famous Incompleteness Theorem, published in 1931, shows that no consistent system of axioms can prove every true arithmetical statement. In particular, it cannot prove an arithmetized version of its own consistency statement. Turing translated this phenomenon into his theorem about computation—that the Halting Problem for an idealized model of computation is undecidable.

Given these stupendous results, it is comparatively trivial to produce an unsettleable Collatzian game. In a 1972 paper, "Unpredictable Iterations" [1], I showed that any computation can be simulated by a Collatzian game of a very simple type, namely a *fraction game*, where the multipartite linear function involved has the form $r_1 n | r_2 n | \ldots r_k n$ determined by a sequence $r_1, r_2, \ldots, r_n$ of rational numbers. The later paper "Fractran: A Simple Universal Programming Language for Arithmetic" [2], shows that the game whose fractions are

$$\frac{583}{559} \; \frac{629}{551} \; \frac{437}{527} \; \frac{82}{517} \; \frac{615}{329} \; \frac{371}{129} \; \frac{1}{115} \; \frac{53}{86} \; \frac{43}{53} \; \frac{23}{47} \; \frac{341}{46} \; \frac{41}{43}$$

$$\frac{47}{41} \; \frac{29}{37} \; \frac{37}{31} \; \frac{299}{29} \; \frac{47}{23} \; \frac{161}{15} \; \frac{527}{19} \; \frac{159}{7} \; \frac{1}{17} \; \frac{1}{13} \; \frac{1}{3}$$

is universal in the sense that for any computable (technically, general recursive) function $f(n)$, there is a constant c such that the game takes $c \cdot 2^{2^n}$ to $2^{2^{f(n)}}$. In this case, we define $f_c(n)$ to be $f(n)$. Moreover, the result includes all partial recursive functions (those that are not always defined) when we say that $f_c(n)$ is undefined if this game does not stop or stops at a number not of the form 2^{2^m}.

From this it follows fairly easily that whatever consistent axioms we use to define "settleable," there is some number for which the game with one more fraction,

$$\frac{583}{559} \; \frac{629}{551} \; \frac{437}{527} \; \frac{82}{517} \; \frac{615}{329} \; \frac{371}{129} \; \frac{1}{115} \; \frac{53}{86} \; \frac{43}{53} \; \frac{23}{47} \; \frac{341}{46} \; \frac{41}{43}$$

$$\frac{47}{41} \; \frac{29}{37} \; \frac{37}{31} \; \frac{299}{29} \; \frac{47}{23} \; \frac{161}{15} \; \frac{527}{19} \; \frac{159}{7} \; \frac{1}{17} \; \frac{1}{13} \; \frac{1}{3} \; \frac{1}{2}$$

never gets to 1, but this is not settleable. Instructions to the writer of a computer program: If the machine succeeds in proving $0 = 1$ from the nth axiom system, define $f(n) = 0$, otherwise leave $f(n)$ undefined. Then, precisely when the system is inconsistent, the 23-fraction game stops at 2, since 0 is the only possible value for $f(n)$, and so the 24-fraction one stops at 1.

What are the simplest Collatzian games that we can expect to be unsettleable? I think I have one answer.

The Amusical Permutation

The *amusical permutation* $\mu(n)$ maps $2k \mapsto 3k$, $4k + 1 \mapsto 3k + 1$, and $4k - 1 \mapsto 3k - 1$. This is obviously a tripartite linear function, since every number is uniquely of one of the three forms on the left-hand side. Since every number is also uniquely of one of the forms on the right-hand side, μ^{-1} is equally a tripartite linear function, and so μ is a permutation. In the abbreviated notation, the amusical permutation is $\frac{3n/2}{2} \mid \frac{3n+1}{4} \mid \frac{3n-1}{4}$, and its inverse is $\frac{2n}{3} \mid \frac{4n+1}{3} \mid \frac{4n-1}{3}$. Using $\{r\}$ for the nearest integer to r, we could abbreviate the permutation μ still further to $\frac{3n}{2} \mid \{\frac{3n}{4}\}$ and μ^{-1} to $\frac{2n}{3} \mid \{\frac{4n}{3}\}$, but this might obscure the fact that μ and μ^{-1} are tripartite rather than bipartite linear functions.

In the usual cycle notation (including possibly infinite cycles), μ begins

$$(\mathbf{1}) \qquad (\mathbf{2}, 3) \qquad (\mathbf{4}, 6, 9, 7, 5)$$
$$(\mathbf{44}, 66, 99, 74, 111, 83, 62, 93, 70, 105, 79, 59)$$

$$(\ldots, 91, 68, \ldots, 86, \ldots, 97, 73, 55, 41, 31, 23, 17, 13, 10, 15,$$
$$11, \mathbf{8}, 12, 18, 27, 20, 30, 45, 34, 51, 38, 57, 43, 32, 48, 72, \ldots)$$

$$(\ldots, 77, 58, 87, 65, 49, 37, 28, 42, 63, 47, 35, 26, 39, 29,$$
$$22, 33, 25, 19, \mathbf{14}, 21, 16, 24, 36, 54, 81, 61, 46, 69, 52,$$
$$78, \ldots, 88, \ldots, 94, \ldots, 89, 67, 50, 75, 56, 84, \ldots)$$

$$(\ldots, 98, \ldots, 100, \ldots, 95, 71, 53, \mathbf{40},$$
$$60, 90, \ldots, 76, \ldots)$$

$$(\ldots, 85, \mathbf{64}, 96, \ldots) \qquad (\ldots, \mathbf{80}, \ldots)$$
$$(\ldots, 92, \ldots, \mathbf{82}, \ldots)$$

wherein the smallest element in each cycle is bold. I have shown what seem to be all the finite cycles and the first six infinite ones, so as to include all numbers up to 100.

Strictly speaking, I do not know that these statements are true. For instance, the cycle containing 8 might be finite, or might be the same as the one containing 14. However, the numbers in both of these cycles have been followed in each direction until they get larger than 10^{400}, and it's obvious that they will never again descend below 100. We need a name for this kind of obviousness: I suggest *probvious*, abbreviating "probabilistically obvious."

Figure 1 makes this point even more clear. It shows the cycles containing 8, 14, 40, 64, 80, and 82 on a logarithmic scale against applications of μ. These six curves have been separated since on the scale displayed their least points are all indistinguishable from 1.[*] The spots indicate where they pass 10^{400}. In both directions the growth is exponential, and μ has a slightly faster rate than μ^{-1}. How can these facts be explained?

Let's consider what is probably the case when the numbers get large. Since a number n is equally likely to be even or odd, it will be multiplied on average by $\sqrt{3/2 \times 3/4} = \sqrt{9/8}$ per move. In twelve moves, the expected factor is

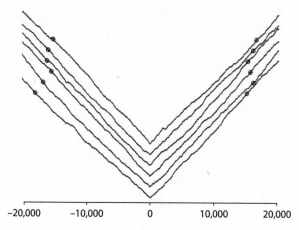

-20,000   -10,000   0   10,000   20,000

FIGURE 1. Cycles from 8, 14, 40, 64, 80, and 82 for 20,000 iterations.

$$\frac{3^{12}}{2^{18}} = \frac{531\ 441}{262\ 144} \approx 2.027.$$

For μ^{-1} we multiply by 2/3 in one case out of three and 4/3 in the other two cases, so the expected increase in three moves is 32/27, and the expected increase in twelve moves is

$$\frac{32^4}{27^4} = \frac{2^{20}}{3^{12}} = \frac{1\ 048\ 576}{531\ 441} \approx 1.973.$$

Taking these two numbers to be 2 explains the name "amusical." On a piano, there are 12 notes per octave, which represents a doubling of frequency, just as 12 steps of the amusical permutation approximately doubles a number, on average. A frequency ratio of

$$\frac{3^{12}}{2^{19}} = \frac{531\ 441}{524\ 288} \approx 1.0136$$

is called the "Pythagorean comma" and is the ratio between B-flat and A-sharp and other pairs of "enharmonically equivalent" notes. So there really is a connection with music. However, since the series always ascends by a fifth modulo octaves, it does not sound very musical, and it has amused me to call it amusical.

Amusical Unsettleabilities?

The simplest assertion about μ that I believe to be true but unsettleable is that 8 belongs to an infinite cycle.

Why is this true? Because the assertion that the logarithm of $\mu^n(8)$ increases linearly is amply verified by Figure 1, and nobody can seriously believe that $\mu^n(8)$, having already surpassed 10^{400}, will miraculously decrease to 8 again (Figure 2, produced after this text was written, shows that after 200,000 iterations it even surpasses 10^{5000}). Being true, the assertion will not be disprovable.

If a Collatzian game does not terminate, is there a proof that it does not terminate? The 24-fraction game in the section called "Are There Unsettleable Collatzian Games?" (which was improved to 7 fractions by John Rickard [3]) shows that in general the answer is no. In general, if a Collatzian game does not stop, then there is no proof of this. So one should not expect the cycle of 8 to be provably infinite in the absence of any reason why it should be. After all, there is a very small positive probability that for some very large positive numbers M and N, $\mu^M(8)$ might just happen to be the same googol digit number as $\mu^{-N}(8)$.

Some readers will still be disappointed not to be given proofs, despite the warning in the introduction that this is clearly impossible. I

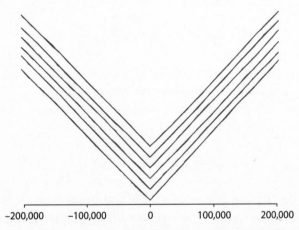

-200,000   -100,000   0   100,000   200,000

FIGURE 2. The same cycles for 200,000 iterations, showing much greater regularity in the long run, and further confirming the probabilistic predictions.

leave such readers with the intriguing thought that the proportion of fallacies in published proofs is far greater than the small positive probability mentioned in the previous paragraph.

Appendix 1: Is the 3n + 1 Problem Settleable?

The $3n + 1$ game presents special features, in that the probabilistic arguments suggest that large numbers decrease, rather than increase as in the amusical permutation. If this argument were provable, the conjecture would be settled by being provable. There is some slight hope that this proof might happen. The celebrated Hardy-Littlewood circle method often makes it possible to prove results that are predicted probabilistically.

Its most spectacular application has been Vinogradov's proof that every sufficiently large odd number is the sum of 3 primes. The method applies more generally to find the number of representations of a number n as a sum of a given number of numbers of some special form (primes, kth-powers, . . .). Their estimate for this number takes the form $P + E$, where P is a probabilistic estimate and E an error term. One hopes to prove that $|E| < P$ so that there is a representation.

P turns out to be a product containing factors P_p, where P_p (for prime p) is the probability that n is p-adically (i.e., modulo all powers of p) the sum of the given number of numbers of that form. (There is also a factor P_∞, which is the proportion of numbers near to N that are representable.) In other words, P is just what one would naively expect from probabilistic considerations analogous to the ones we used for the amusical permutation.

It is not entirely inconceivable that such a method might one day prove the Collatz $3n + 1$ Conjecture, since all one has to do is prove that large enough numbers eventually reduce. However, I don't really believe it.

These remarks do not apply to the amusical permutation, whose behavior would not be established even if one proved that almost all large numbers tend to increase, since, for instance, the number obtained by applying μ a million times to 8 might just be the same as the number obtained by applying μ^{-1} rather more times to 8 or 14, in which case the cycle containing 8 would be either finite or the same as the cycle containing 14. This probviously doesn't happen, but we can't expect to prove it, and there's no reason to expect that either it or its negative follows the Zermelo-Fraenkel Axioms or any likely extension of them. In other words, it's probviously unsettleable.

Some other things are probvious but with a slightly smaller probability. For instance, there is probviously an algorithm for telling whether n belongs to a finite cycle. Just ask whether n is one of the twenty numbers:

$$1, 2, 3, 4, 5, 6, 7, 9, 44, 59, 62, 66, 70, 74, 79, 83, 93, 99, 105, 111;$$

if so, say "yes," if not, "no." If there is another finite cycle this algorithm fails, but the answer will still be computable unless there are infinitely many finite cycles, which there very probviously aren't.

I've already suggested that the assertion that 8 is in an infinite cycle, although probvious, is unsettleable. I now propose that this unsettleability assertion is itself unprovable, and therefore unsettleable, and so on arbitrarily far into the metatheory.

Even if this is wrong, mathematics is *not* defined by any system of set theoretical axioms. In particular, it is likely that some simple Collatzian problems (possibly even the $3n + 1$ problem itself) will remain forever unsettleable.

Appendix 2: Some Amusical Paradoxes

With some relief, let's put deep problems aside to discuss some simple puzzles about the behavior of the amusical permutation. We have already noticed the "either-way-up paradox," that the numbers in the typical cycle increase no matter which way we move along the cycle. It's not really paradoxical, as Figure 1 shows. No matter where we start on the cycle and no matter which direction we move, we'll eventually pass the minimum, and after that we go up.

Here is the "Congruence Paradox." Since $n < \mu(n)$ only if n is even and $n < \mu^{-1}(n)$ only if n is not a multiple of 3, it satisfies both these inequalities (and so is a local minimum) only if $n \equiv \pm 2 \pmod 6$, which happens in exactly one third of cases: Right? Maybe not. It satisfies neither inequality only if $n \equiv 3 \pmod 6$ and so is a local maximum in exactly one sixth of the cases. But in any sequence, local minima and maxima alternate, so there should be just as many of each. So which is right: Do we get these turning points every third term or every sixth term?

Let's think again. Whenever an increase is followed by a decrease we get a maximum, and since increases and decreases are equally likely, we should get a maximum one quarter of the time (and the same argument applies to minima), which happen when a decrease is followed by an increase. So these things both happen once in four moves rather than once in either three or six! We can get yet another answer by thinking backwards, when the two probabilities are 2/3 and 1/3, leading to the conclusion that maxima and minima both occur once every 4 1/2 moves.

What these arguments prove is not really paradoxical. If one follows a typical number, one sees both maxima and minima equally often, namely, once every four moves going forward or once every 4 1/2 backward. We leave it to the reader to explain why neither of these answers (once in 4 or 4 1/2) agrees with either of the answers (once in 3 or 6) given by the Congruence Paradox.

Since the apparent contradictions are based on our experience with finite cycles, one might think that they could be turned around to prove that most cycles are infinite, or that at least there are some infinite cycles. However, having thought about it, I still believe that these problems are unsettleable.

If you disagree, try to prove or disprove either of the following statements:

1. There is a new finite cycle.
2. There is an infinite cycle.

Acknowledgments

Alex Ryba deserves many thanks for his invaluable help in producing this paper. I would also like to thank Dierk Schleicher for having produced the pictures.

Postscript

Added June 8, 2012. The following argument has convinced me that the Collatz $3n + 1$ Conjecture is itself very likely to be unsettleable, rather than this merely having the slight chance mentioned above. It uses the fact that there are arbitrarily tall "mountains" in the graph of the Collatz game. To see this, observe that $2m - 1$ passes in two moves to $3m - 1$, from which it follows that $2^k m - 1$ passes in $2k$ moves to $3^k m - 1$. Now by the Chinese Remainder Theorem, we can arrange that

$3^k m - 1$ has the form $2^l n$, which passes by l moves to n. There is a very slight possibility that n happens to be the same as the number $2^k m - 1$ that we started with. Let's suppose that the starting number $2^k m - 1$ is about a googol; then the downward slope of the mountain certainly contains a number between one and two googols, so the chance that this is the same as the starting number is at least one googolth. (This result is justified by observations for smaller n showing that the first iterate that lies in the range $[n, 2n)$ is approximately uniformly distributed in this range.) In my view, the fact that this probability, though very small, is positive, makes it extremely unlikely that there can be a proof that the Collatz game has no cycles that contain only large numbers. This should not be confused with a suggestion that there actually *are* cycles containing large numbers. After all, events whose probability is around one googolth are distinctly unlikely to happen!

I don't want readers to take these words on trust but rather to encourage those who don't find them convincing to try even harder to prove the Collatz Conjecture!

Notes

* The visible kink in the graph of the 82 cycle corresponds to the remarkable decrease (by a factor of more than 75,989) from $\mu^{1981}(82) = 5\,518\,82\,09\,452\,689\,749\,562\,442\,051\,558\,599\,474\,342\,616\,171\,049\,802\,024\,438\,847\,761 \approx 5.519 \times 10^{63}$ to $\mu^{2208}(82) = 72\,625\,599\,594\,039\,327\,995\,887\,556\,149\,205\,597\,399\,175\,812\,389\,461\,574\,936\,396 \approx 7.263 \times 10^{58}$. Admittedly, this decrease by a factor of more than 2^{16} where an increase of almost 2^{19} was to be expected casts some doubt on the probabilistic arguments in the text.

References

1. J. H. Conway, "Unpredictable Iterations." *Proceedings of the Number Theory Conference* (University of Colorado, Boulder, CO, 1972), pp. 49–52. Also available in *The Ultimate Challenge: The 3x + 1 Problem*, Edited by Jeffrey C. Lagarias, American Mathematical Society, Providence, RI, 2010.

2. J. H. Conway, "Fractran: A Simple Universal Programming Language for Arithmetic." Ch. 2 in *Open Problems in Communication and Computation*, Edited by T. M. Cover and B. Gopinath, Springer-Verlag, New York, 1987, pp. 4–26. Also available in *The Ultimate Challenge: The 3x+1 Problem*, Edited by Jeffrey C. Lagarias, American Mathematical Society, Providence, RI, 2010.

3. J. Rickard, unpublished. (John Rickard died May 9, 2002.)

4. T. Oliveira e Silva, "Empirical Verification of the 3x + 1 and Related Conjectures." In *The Ultimate Challenge: The 3x + 1 Problem*, Edited by Jeffrey C. Lagarias, American Mathematical Society, Providence, RI, 2010, pp. 189–207.

FIGURE 1 from "The Music of Math Games." (Image courtesy of BrainQuake, Inc.)

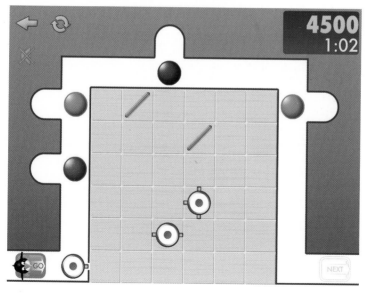

FIGURE 2 from "The Music of Math Games." (Screenshot courtesy of MIND Research Institute.)

FIGURE 3 from "The Music of Math Games." (Image courtesy of Motion Math.)

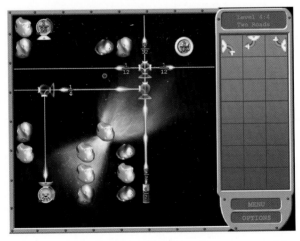

FIGURE 4 from "The Music of Math Games." (Image courtesy of the Center for Game Science, University of Washington.)

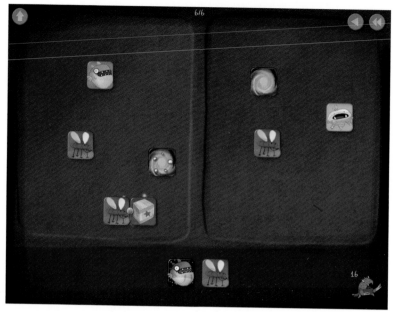

FIGURE 5 from "The Music of Math Games." (Image courtesy of WeWantToKnow AS.)

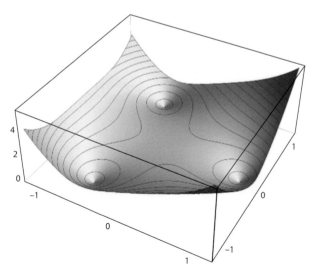

FIGURE 1 from "The Fundamental Theorem of Algebra for Artists."

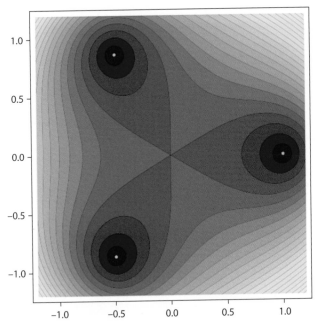

FIGURE 2 from "The Fundamental Theorem of Algebra for Artists."

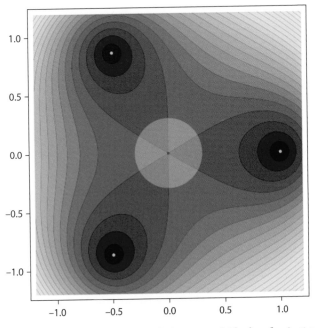

FIGURE 3 from "The Fundamental Theorem of Algebra for Artists."

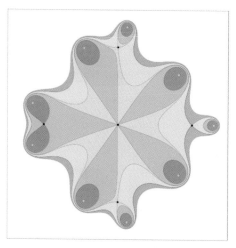

FIGURE 5 from "The Fundamental Theorem of Algebra for Artists."

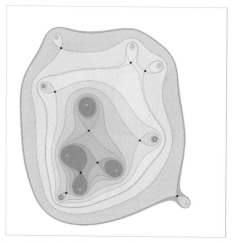

FIGURE 6 from "The Fundamental Theorem of Algebra for Artists."

FIGURE 3 from "On the Number of Klein Bottle Types."

FIGURE 4 from "On the Number of Klein Bottle Types."

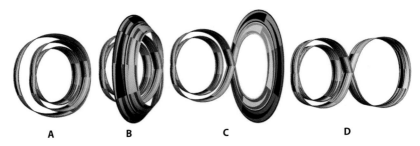

FIGURE 5 from "On the Number of Klein Bottle Types."

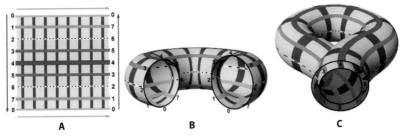

FIGURE 6 from "On the Number of Klein Bottle Types."

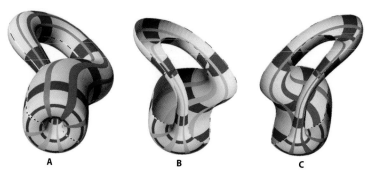

FIGURE 7 from "On the Number of Klein Bottle Types."

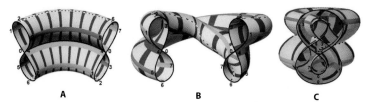

FIGURE 8 from "On the Number of Klein Bottle Types."

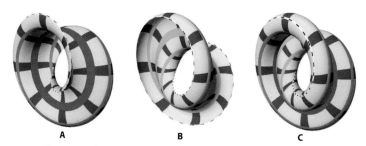

FIGURE 9 from "On the Number of Klein Bottle Types."

FIGURE 10 from "On the Number of Klein Bottle Types."

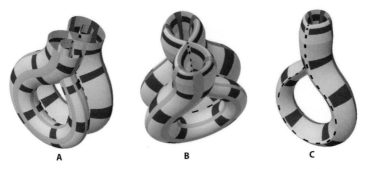

FIGURE 11 from "On the Number of Klein Bottle Types."

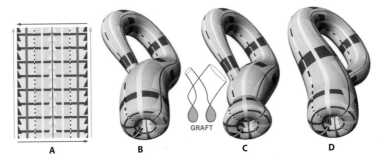

FIGURE 12 from "On the Number of Klein Bottle Types."

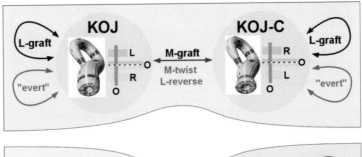

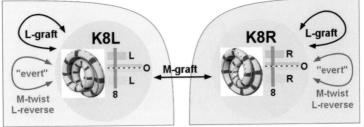

FIGURE 13 from "On the Number of Klein Bottle Types."

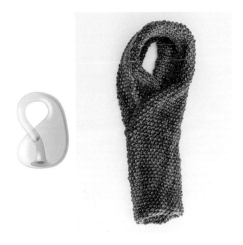

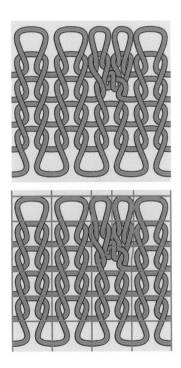
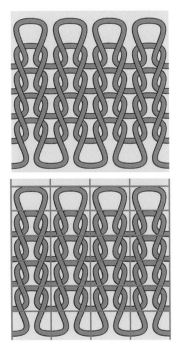

Figure 1 from "Adventures in Mathematical Knitting." Knitted object made by sarah-marie belcastro. Photograph by Austin Green. All illustrations from "Adventures in Mathematical Knitting" courtesy of Scientific American.

Figure 2 from "Adventures in Mathematical Knitting."

Figure 3 from "Adventures in Mathematical Knitting."

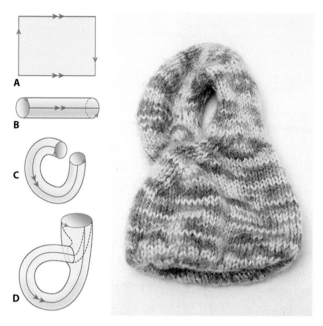

FIGURE 5 from "Adventures in Mathematical Knitting." Knitted object made by sarah-marie belcastro. Photograph by Austin Green.

FIGURE 7 from "Adventures in Mathematical Knitting." Knitted object made by sarah-marie belcastro. Photograph by Austin Green.

FIGURE 8 from "Adventures in Mathematical Knitting." Knitted object made by sarah-marie belcastro. Photograph by Austin Green.

FIGURE 9 from "Adventures in Mathematical Knitting." Knitted object made by sarah-marie belcastro. Photograph by Austin Green.

FIGURE 11 from "Adventures in Mathematical Knitting." Knitted object made by sarah-marie belcastro. Photograph by Austin Green.

FIGURE 2 from "The Mathematics of Fountain Design."

FIGURE 2 from "Chaos at Fifty." (Image courtesy of Stefan Ganev.)

A

B

Figure 3 from "Chaos at Fifty." (Adapted from Tél and Gruiz 2006, color plate VII. Copyright © 2006 Tamás Tél and Márton Gruiz. Reprinted with the permission of Cambridge University Press)

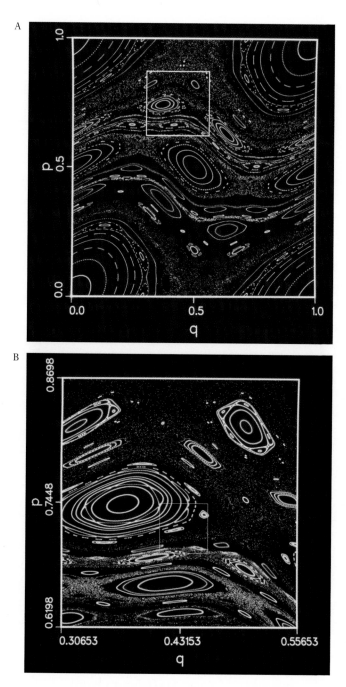

FIGURE 4 from "Chaos at Fifty." (Adapted from Campbell 1989.)

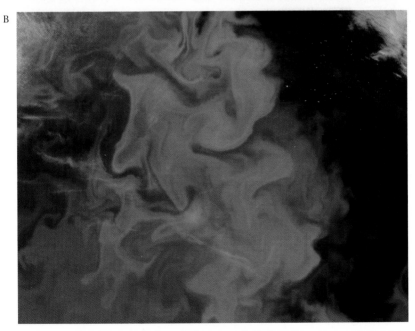

FIGURE 5A and figure 5B from "Chaos at Fifty." (Part A courtesy of NASA/ Cassini Mission; Part B adapted from NASA/Ocean Color Web.)

C

FIGURE 5C from "Chaos at Fifty." (Courtesy of Christophe Gissinger).

Crinkly Curves

Brian Hayes

In 1877, the German mathematician Georg Cantor made a shocking discovery. He found that a two-dimensional surface contains no more points than a one-dimensional line. Cantor compared the set of all points forming the area of a square with the set of points along one of the line segments on the perimeter of the square. He showed that the two sets are the same size. Intuition rebels against this notion. Inside a square you could draw infinitely many parallel line segments side by side. Surely an area with room for such an infinite array of lines must include more points than a single line—but it doesn't. Cantor himself was incredulous: "I see it, but I don't believe it," he wrote.

Yet the fact was inescapable. Cantor defined a one-to-one corre-spondence between the points of the square and the points of the line segment. Every point in the square was associated with a single point in the segment; every point in the segment was matched with a unique point in the square. No points were left over or used twice. It was like pairing up mittens: If you come out even at the end, you must have started with equal numbers of lefts and rights.

Geometrically, Cantor's one-to-one mapping is a scrambled affair. Neighboring points on the line scatter to widely separated destinations in the square. The question soon arose: Is there a *continuous* mapping between a line and a surface? In other words, can one trace a path through a square without ever lifting the pencil from the paper and touch every point at least once? It took a decade to find the first such curve. Then dozens more were invented, as well as curves that fill up a three-dimensional volume or even a region of some n-dimensional space. The very concept of dimension was undermined.

Circa 1900, these space-filling curves were viewed as mysterious aberrations, signaling how far mathematics had strayed from the world of everyday experience. The mystery has never entirely faded away, but the curves have grown more familiar. They are playthings of programmers now, nicely adapted to illustrating certain algorithmic techniques (especially recursion). More surprising, the curves have turned out to have practical applications. They serve to encode geographic information; they have a role in image processing; they help allocate resources in large computing tasks. And they tickle the eye of those with a taste for intricate geometric patterns.

How to Fill Up Space

It's easy to sketch a curve that completely fills the interior of a square. The finished product looks like this:

How uninformative! It's not enough to know that every point is covered by the passage of the curve; we want to see how the curve is constructed and what route it follows through the square.

If you were designing such a route, you might start out with the kind of path that's good for mowing a lawn:

But there's a problem with these zigzags and spirals. A mathematical lawn mower cuts a vanishingly narrow swath, and so you have to keep reducing the space between successive passes. Unfortunately, the limiting pattern when the spacing goes to zero is not a filled area; it is a path that forever retraces the same line along one edge of the square or around its perimeter, leaving the interior blank.

The first successful recipe for a space-filling curve was formulated in 1890 by Giuseppe Peano, an Italian mathematician also noted for his axioms of arithmetic. Peano did not provide a diagram or even an explicit description of what his curve might look like; he merely defined a pair of mathematical functions that give x and y coordinates inside a square for each position t along a line segment.

Soon David Hilbert, a leading light of German mathematics in that era, devised a simplified version of Peano's curve and discussed its

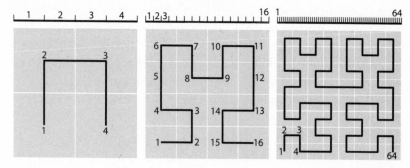

FIGURE 1. A space-filling curve evolves through successive stages of refinement as it grows to cover the area of a square. This illustration is a redrawing of the first published diagram of such a curve; the original appeared in an 1891 paper by David Hilbert. The idea behind the construction is to divide a line segment into four intervals and divide a square into four quadrants, then establish a correspondence between the points of corresponding intervals and quadrants. The process continues with further recursive subdivisions.

geometry. Figure 1 is a redrawing of a diagram from Hilbert's 1891 paper, showing the first three stages in the construction of the curve.

Programming by Procrastination

Figure 2 shows a later stage in the evolution of the Hilbert curve, when it has become convoluted enough that one might begin to believe it will eventually reach all points in the square. The curve was drawn by a computer program written in a recursive style that I call programming by procrastination. The philosophy behind the approach is this: Plotting all those twisty turns looks like a tedious job, so why not put it off as long as we can? Maybe we'll never have to face it.

Let us eavesdrop on a computer program named *Hilbert* as it mumbles to itself while trying to solve this problem:

Hmm. I'm supposed to draw a curve that fills a square. I don't know how to do that, but maybe I can cut the problem down to size. Suppose I had a subroutine that would fill a smaller square, say one-fourth as large. I could invoke that procedure on each quadrant of the main square, getting back four separate pieces of the space-filling curve. Then, if I just draw three line segments to link the four pieces into one long curve, I'll be finished!

Of course, I don't actually have a subroutine for filling in a quadrant. But a quadrant of a square is itself a square. There's a program named *Hilbert* that's supposed to be able to draw a space-filling curve in any square. I'll just hand each of the quadrants off to *Hilbert*.

FIGURE 2. After seven stages of elaboration, the Hilbert curve meanders through $4^7 = 16,384$ subdivisions of the square. The curve is an unbranched path with end points at the lower left and lower right. It is continuous in the sense that it has no gaps or jumps, but it is not smooth: All of the right angles are points where the curve has no tangent (or, in terms of calculus, no derivative). Continuing the subdivision process leads to a limiting case where the curve fills the entire square, showing that a two-dimensional square has no more points than a one-dimensional line segment.

The strategy described in this monologue may sound like a totally pointless exercise. The *Hilbert* program keeps subdividing the problem but has no plan for ever actually solving it. However, this is one of those rare and wonderful occasions when procrastination pays off, and the homework assignment you lazily set aside last night is miraculously finished when you get up in the morning.

Consider the sizes of the successive subsquares in *Hilbert*'s divide-and-conquer process. At each stage, the side length of the square is halved, and the area is reduced to one-fourth. The limiting case, if the process goes on indefinitely, is a square of zero side length and zero area. So here's the procrastinator's miracle: Tracing a curve that touches all the points inside a size-zero square is easy because such a square is in fact a single point. Just draw it!

Practical-minded readers will object that a program running in a finite machine for a finite time will not actually reach the limiting case of squares that shrink away to zero size. I concede the point. If the recursion is halted while the squares still contain multiple points, one of those points must be chosen as a representative; the center of the square is a likely candidate. In making the illustration above, I stopped the program after seven levels of recursion, when the squares were small but certainly larger than a single point. The wiggly line connects the centers of $4^7 = 16,384$ squares. Only in the mind's eye do we ever see a true, infinite space-filling curve, but a finite drawing like this one is at least a guide to the imagination.

There is one more important aspect of this algorithm that I have glossed over. If the curve is to be continuous—with no abrupt jumps—then all the squares have to be arranged so that one segment of the curve ends where the next segment begins. Matching up the end points in this way requires rotating and reflecting some of the subsquares. (For an animated illustration of these geometric transformations, see http://bit-player.org/extras/hilbert.)

Grammar and Arithmetic

The procrastinator's algorithm is certainly not the only way to draw a space-filling curve. Another method exploits the self-similarity of the pattern—the presence of repeated motifs that appear in each successive stage of the construction. In the Hilbert curve, the basic motif is a U-shaped path with four possible orientations. In going from one stage

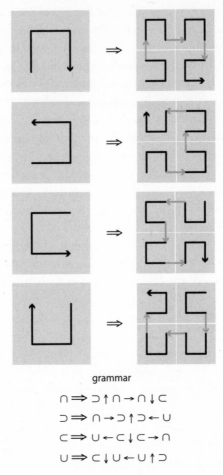

grammar

∩ ⇒ ⊃ ↑ ∩ → ∩ ↓ ⊂
⊃ ⇒ ∩ → ⊃ ↑ ⊃ ← ∪
⊂ ⇒ ∪ ← ⊂ ↓ ⊂ → ∩
∪ ⇒ ⊂ ↓ ∪ ← ∪ ↑ ⊃

FIGURE 3. Substitution rules generate the Hilbert curve by replacing a U-shaped motif in any of four orientations with sequences of four rotated and reflected copies of the same motif. The rules constitute a generative grammar. An interactive version of the illustration is available online at http://bit-player.org/extras/hilbert.

of refinement to the next, each U orientation is replaced by a specific sequence of four smaller U curves, along with line segments that link them together, as shown in Figure 3. The substitution rules form a grammar that generates geometric figures in the same way that a linguistic grammar generates phrases and sentences.

The output of the grammatical process is a sequence of symbols. An easy way to turn it into a drawing is to interpret the symbols as commands in the language of "turtle graphics." The turtle is a conceptual drawing instrument, which crawls over the plane in response to simple instructions to move forward, turn left, or turn right. The turtle's trail across the surface becomes the curve to be drawn.

When Peano and Hilbert were writing about the first space-filling curves, they did not explain them in terms of grammatical rules or turtle graphics. Instead their approach was numerical, assigning a number in the interval [0, 1] to every point on a line segment and also to every point in a square. For the Hilbert curve, it's convenient to do this arithmetic in base 4, or quaternary, working with the digits 0, 1, 2, and 3. In a quaternary fraction such as 0.213, each successive digit specifies a quadrant or subquadrant of the square, as outlined in Figure 4.

What about other space-filling curves? Peano's curve is conceptually similar to Hilbert's but divides the square into nine regions instead

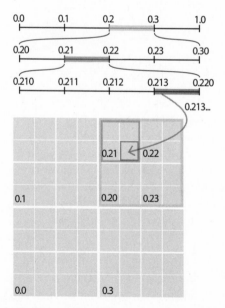

FIGURE 4. Base-4 encoding of the Hilbert curve shows how fourfold divisions of the unit interval [0, 1] are mapped onto quadrants of the square. For example, any base-4 number beginning 0.213 must lie in the smallest of the squares outlined in shades of gray.

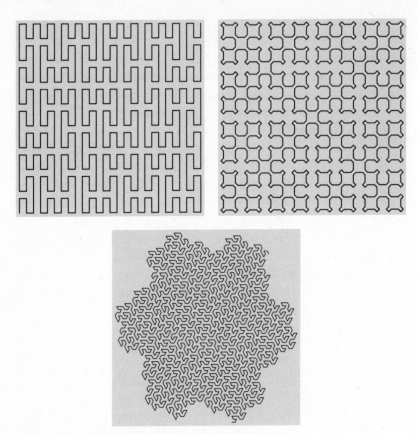

FIGURE 5. The first space-filling curve (top left) was described in 1890 by
Italian mathematician Giuseppe Peano; the construction divides a square
into nine smaller squares. A curve based on a triangular dissection (top right)
was introduced in 1912 by Polish mathematician Wacław Sierpiński. The
"flowsnake" curve (bottom), invented by American mathematician Bill Gosper
in the 1970s, fills a ragged-edged hexagonal area.

of four. Another famous example was invented in 1912 by the Polish
mathematician Wacław Sierpiński; it partitions the square along its di-
agonals, forming triangles that are then further subdivided. A more
recent invention is the "flowsnake" curve devised in the 1970s by Bill
Gosper.

Filling three-dimensional space turns out to be even easier than
filling the plane—or at least there are more ways to do it. Herman

Haverkort of the Eindhoven Institute of Technology in the Netherlands has counted the three-dimensional analogues of the Hilbert curve; there are more than 10 million of them.

All Elbows

In everyday speech, the word *curve* suggests something smooth and fluid, without sharp corners, such as a parabola or a circle. The Hilbert curve is anything but smooth. All finite versions of the curve consist of 90-degree bends connected by straight segments. In the infinite limit, the straight segments dwindle away to zero length, leaving nothing but sharp corners. The curve is all elbows. In 1900, the American mathematician Eliakim Hastings Moore came up with the term "crinkly curves" for such objects.

In many respects, these curves are reminiscent of fractals, the objects of fractional dimension that Benoit Mandelbrot made famous. The curves' self-similarity is fractal-like: Zooming in reveals ever more intricate detail. But the Hilbert curve is not a fractal because its dimension is not a fraction. Any finite approximation is simply a one-dimensional line. On passing to the limit of infinite crinkliness, the curve suddenly becomes a two-dimensional square. There is no intermediate state.

Even though the complete path of an infinite space-filling curve cannot be drawn on paper, it is still a perfectly well-defined object. You can calculate the location along the curve of any specific point you might care to know about. The result is exact if the input is exact. A few landmark points for the Hilbert curve are plotted in Figure 6.

The algorithm for this calculation implements the definition of the curve as a mapping from a one-dimensional line segment to a two-dimensional square. The input to the function is a number in the interval [0, 1], and the output is a pair of x, y coordinates.

The inverse mapping—from x, y coordinates to the segment [0, 1]— is more troublesome. The problem is that a point in the square can be linked to more than one point on the line.

Cantor's dimension-defying function was a *one-to-one* mapping: Each point on the line was associated with exactly one point in the square, and vice versa. But Cantor's mapping was not continuous: Adjacent points on the line did not necessarily map to adjacent points in the

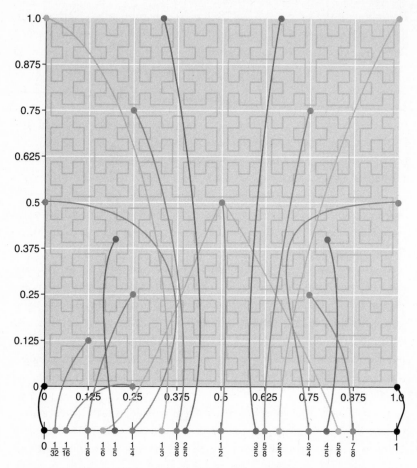

FIGURE 6. Positions of points along the infinitely crinkled course of the Hilbert curve can be calculated exactly, even though the curve itself cannot be drawn. Here 19 selected points in the interval [0, 1] are mapped to coordinates in the unit square, $[0, 1]^2$. Although a finite approximation to the Hilbert curve is shown in the background, the positions within the square are those along the completed, infinite curve. The inverse mapping is not unique: Points in the square map back to multiple points in the interval. An interactive version of the illustration is available online at http://bit-player.org /extras/hilbert/.

square. In contrast, the space-filling curves are continuous but not one-to-one. Although each point on the line is associated with a unique point in the square, a point in the square can map back to multiple points on the line. A conspicuous example is the center of the square, with the coordinates $x = 1/2$, $y = 1/2$. Three separate locations on the line segment (1/6, 1/2, and 5/6) all connect to this one point in the square.

Math on Wheels

Space-filling curves have been called monsters, but they are useful monsters. One of their most remarkable applications was reported 30 years ago by John J. Bartholdi III and his colleagues at the Georgia Institute of Technology. Their aim was to find efficient routes for drivers delivering Meals on Wheels to elderly clients scattered around the city of Atlanta. Finding the best possible delivery sequence would be a challenging task even with a powerful computer. Meals on Wheels didn't need the solution to be strictly optimal, but they needed to plan and revise routes quickly, and they had to do it with no computing hardware at all. Bartholdi and his coworkers came up with a scheme that used a map, a few pages of printed tables, and two Rolodex files.

Planning a route started with Rolodex cards listing the delivery addresses. The manager looked up the map coordinates of each address, then looked up those coordinates in a table, which supplied an index number to write on the Rolodex card. Sorting the cards by index number yielded the delivery sequence.

Behind the scenes in this procedure was a space-filling curve (specifically, a finite approximation to a Sierpiński curve) that had been superimposed on the map. The index numbers in the tables encoded position along this curve. The delivery route didn't follow the Sierpiński curve, with all its crinkly turns. The curve merely determined the sequence of addresses, and the driver then chose the shortest point-to-point route between them.

A space-filling curve works well in this role because it preserves "locality." If two points are nearby on the plane, they are likely to be nearby on the curve as well. The route makes no wasteful excursions across town and back again.

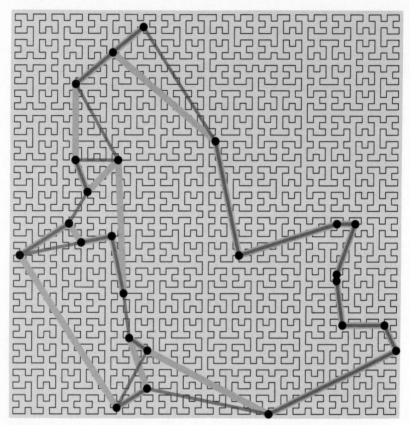

FIGURE 7. Approximate solutions to the traveling salesman problem emerge from a simple algorithm based on space-filling curves. Here 25 cities (black dots) are randomly distributed within a square. The traveling salesman problem calls for the shortest tour that passes through all the cities and returns to the starting point. Listing the cities in the order they are visited by a space-filling curve yields a path of length 274 (thick, light-gray line); the optimal tour (darker gray line) is about 13 percent better, with a length of 239. The space-filling curve used in this example was invented by E. H. Moore in 1900; it is related to the Hilbert curve but forms a closed circuit. The unit of distance for measuring tours is the step size of the Moore curve. The optimal tour was computed with the Concorde TSP Solver (http://www.cs.sunysb .edu/~algorith/implement/concorde/implement.shtml).

The Meals on Wheels scheduling task is an instance of the traveling salesman problem, a notorious stumper in computer science. The Bartholdi algorithm gives a solution that is not guaranteed to be best but is usually good. For randomly distributed locations, the tours average about 25 percent longer than the optimum. Other heuristic methods can beat this performance, but they are much more complicated. The Bartholdi method finds a route without even computing the distances between sites.

Locality is a helpful property in other contexts as well. Sometimes what's needed is not a route from one site to the next but a grouping of sites into clusters. In two or more dimensions, identifying clusters can be difficult; threading a space-filling curve through the data set reduces it to a one-dimensional problem.

The graphic arts have enlisted the help of space-filling curves for a process known as halftoning, which allows black-and-white devices (such as laser printers) to reproduce shades of gray. Conventional halftoning methods rely on arrays of dots that vary in size to represent lighter and darker regions. Both random and regular arrays tend to blur fine features and sharp lines in an image. A halftone pattern that groups the dots along the path of a Hilbert or Peano curve can provide smooth tonal gradients while preserving crisp details.

Another application comes from a quite different realm: the multiplication of matrices (a critical step in large-scale computations). Accessing matrix elements by rows and columns requires the same values to be read from memory multiple times. In 2006, Michael Bader and Christoph Zenger of the Technical University of Munich showed that clustering the data with a space-filling curve reduces memory traffic.

Bader is also the author of an excellent recent book that discusses space-filling curves from a computational point of view. An earlier volume by Hans Sagan is more mathematical.

Given that people have found such a surprising variety of uses for these curious curves, I can't help wondering whether nature has also put them to work. Other kinds of patterns are everywhere in the natural world: stripes, spots, spirals, and many kinds of branching structures. But I can't recall seeing a Peano curve on the landscape. The closest I can come are certain trace fossils (preserved furrows and burrows of organisms on the sea floor) and perhaps the ridges and grooves on the surface of the human cerebrum.

Cantor's Conundrums

Applications of space-filling curves are necessarily built on finite examples—paths one can draw with a pencil or a computer. But in pure mathematics, the focus is on the infinite case, where a line gets so incredibly crinkly that it suddenly becomes a plane.

Cantor's work on infinite sets was controversial and divisive in his own time. Leopold Kronecker, who had been one of Cantor's professors in Berlin, later called him "a corrupter of youth" and tried to block publication of the paper on dimension. But Cantor had ardent defenders, too. Hilbert wrote in 1926, "No one shall expel us from the paradise that Cantor has created." Indeed, no one has been evicted. (A few have left of their own volition.)

Cantor's discoveries eventually led to clearer thinking about the nature of continuity and smoothness, concepts at the root of calculus and analysis. The related development of space-filling curves called for a deeper look at the idea of dimension. From the time of Descartes, it was assumed that in d-dimensional space it takes d coordinates to state the location of a point. The Peano and Hilbert curves overturned this principle: A single number can define position on a line, on a plane, in a solid, or even in those 11-dimensional spaces so fashionable in high-energy physics.

At about the same time that Cantor, Peano, and Hilbert were creating their crinkly curves, the English schoolmaster Edwin Abbott was writing his fable *Flatland*, about two-dimensional creatures that dream of popping out of the plane to see the world in 3D. The Flatlanders might be encouraged to learn that mere one-dimensional worms can break through to higher spaces just by wiggling wildly enough.

Bibliography

Bader, M. 2013. *Space-Filling Curves: An Introduction with Applications in Scientific Computing*. Berlin: Springer.

Bartholdi, J. J., III, L. K. Platzman, R. L. Collins, and W. H. Warden III. 1983. A minimal technology routing system for Meals on Wheels. *Interfaces* 13(3): 1–8.

Dauben, J. W. 1979. *Georg Cantor: His Mathematics and Philosophy of the Infinite*. Cambridge, MA: Harvard University Press.

Gardner, M. 1976. Mathematical games: In which "monster" curves force redefinition of the word "curve." *Scientific American* 235: 124–33.

Hilbert, D. 1891. Über die stetige Abbildung einer Linie auf ein Flächenstück. *Mathematische Annalen* 38: 459–60.

Moore, E. H. 1900. On certain crinkly curves. *Transactions of the American Mathematical Society* 1(1): 72–90.

Null, A. 1971. Space-filling curves, or how to waste time with a plotter. *Software: Practice and Experience* 1: 403–10.

Peano, G. 1890. Sur une courbe, qui remplit toute une aire plane. *Mathematische Annalen* 36: 157–60.

Platzman, L. K., and J. J. Bartholdi III. 1989. Spacefilling curves and the planar travelling salesman problem. *Journal of the Association for Computing Machinery* 36: 719–37.

Sagan, H. 1991. Some reflections on the emergence of space-filling curves: The way it could have happened and should have happened, but did not happen. *Journal of the Franklin Institute* 328: 419–30.

Sagan, H. 1994. *Space-Filling Curves*. New York: Springer-Verlag.

Sierpiński, W. 1912. Sur une nouvelle courbe continue qui remplit toute une aire plane. *Bulletin de l'Académie des Sciences de Cracovie*, Série A, 462–78.

Velho, L., and J. de Miranda Gomes. 1991. Digital halftoning with space filling curves. *Computer Graphics* 25(4): 81–90.

Why Do We Perceive Logarithmically?

Lav R. Varshney and John Z. Sun

How We Perceive the World

Whether it is hearing a predator approaching or a family member talking, whether it is perceiving the size of a cartload of chimpanzees or a pride of lions, whether it is seeing a ripe fruit or an explosion, the ability to sense and respond to the natural world is critical to our survival. Indeed, the world is a random place where all intensities of sensory stimuli can arise, from smallest to largest. We must register them all, from the buzz of an insect to the boom of an avalanche or an erupting volcano: our sensory perception must allow us to efficiently represent the statistical distributions of the natural world.

If kindergarten children are asked to point to the correct location for a spoken number word on a line segment labeled with end points 0 and 100, they place smaller numbers closer to 0 and larger numbers closer to 100, but they do not distribute numbers evenly. They give more space on the line to small numbers. The larger ones are crammed into a narrow space at the "100" end. If they are given the numbers 1 to 10, they put 3 at about halfway. In other words, they place numbers in a compressed logarithmic mapping. They perceive numbers logarithmically.

Fourth graders, however, map linearly rather than logarithmically. But if one asks an adult member of the Mundurucu—an Amazonian indigenous group with a reduced numerical lexicon and little formal education—to complete the same task, the logarithmic mapping occurs again. It seems that placing numbers evenly along a line is something that is learned, not innate. Without the intervention of critical educational experiences, the way we perceive numbers is logarithmic (Dehaene 2011).

Number is important to human survival—you would want to know whether one lion is facing you or several. Indeed, it could be argued that perceiving numbers logarithmically rather than linearly could give an evolutionary advantage: it could be more important to know whether it is five lions facing you or three than to know if the deer herd you are chasing contains 100 animals or just 98. But in fact perceptual systems of all kinds display a nonlinear relationship between external stimulus and internal representation. If we double the force on your hand, it will feel like less than double the pressure. If we double the salinity of water, the taste will not be twice as salty. Nonlinear scalings that give greater perceptual resolution to less intense stimuli are ubiquitous across animal species and across sensory modalities: heaviness, pain, warmth, taste, loudness, pitch, brightness, distance, time delay, and color saturation, among others, are all perceived this way. Moreover, these mappings between observable stimulus and our internal perception space—these psychophysical scales and laws—are approximately logarithmic.

With multifarious organisms adapted to a variety of niches, biology is incredibly variable; so why is the same psychophysical law present in so many animals and so many sensory modes? In our work with Grace Wang and Vivek Goyal at MIT (Sun et al. 2012), we have proposed an explanation based on information theory for why sensory perception is the way it is.

The principle that animals are well adapted to their environments, whether in structure or in behavior, has been a powerful method to answer the "why" questions of biology in formal mathematical ways. For example, the principle precisely predicts the facet size of insect compound eyes by balancing resolution and diffraction for natural light; it predicts exactly when mammals shift their gait from walking to trotting to galloping so as to minimize energy cost of movement. We can apply it to the structure of the brain as well: for example, one of us has previously argued that the physical microarchitecture of synapses in the brain is optimal for memory storage capacity per unit volume (Varshney et al. 2006).

The same principle applies to the way that the brain operates—the way that we receive, process, and perceive information from the outside world. If we do it efficiently, we are at an obvious evolutionary advantage, which invites the question: What architecture of information processing within our brains would qualify as "most efficient"?

One answer might be "one that reduces error to a minimum," which invites the further question: What sort of error needs to be thus reduced?

Here we can turn to the nineteenth-century physicist and biologist Ernst Weber. He used to ask how the structure of the nervous system leads to a functional representation of external space. This question led him to the first general results in experimental psychophysics: that the change in stimulus intensity, ΔS, that will be just noticeable, ΔP, is a constant ratio K of the original stimulus intensity S:

$$\Delta P = K \frac{\Delta S}{S}$$

In other words, we do not notice *absolute* changes in stimuli; we notice *relative* changes. This phenomenon leads to an answer to the question above: the error that needs to be reduced in the brains of organisms such as ourselves is not absolute error, but relative error.

By solving Weber's differential equation and broadly putting forth a philosophy that internal representations are tied to properties of the external world, Gustav Fechner developed what is now called the Weber-Fechner law (Fechner 1860). It states that perceived intensity P is logarithmic to the stimulus intensity S (above a minimal threshold of perception S_0)—see box. It has become a centerpiece of psychophysics. But until now it has been an empirical law only. Theory to explain why it should be has been lacking.

Invoking the optimization approach to biology, we can argue that organisms are well adapted to the statistics of natural external stimuli— the distributions of intensities that assail our sensory organs—and are also internally efficient for processing the information that results. This simple principle provides a formal methodology for explaining— indeed for predicting—the logarithmic laws that seem to govern so much of our perception of the world. By applying our optimization approach to reducing the relative error in information processing, we can produce a model that shows why the Weber-Fechner law should apply, and why we perceive so many things logarithmically.

Toward an Optimality Theory

If you are trying to minimize relative error, how should you go about doing it? In other words, how should we go about formulating an optimality theory for sensory perception? Information theory, the statistical theory of communication, was first formulated by Claude Shannon

The Weber-Fechner law states that, above a minimal threshold of perception S_0, perceived intensity P is logarithmic to stimulus intensity S:

$$P = K \log \frac{S}{S_0}$$

Although not precisely true in all sensory regimes—indeed, several alternative psychophysical laws such as Stevens's power law have been proposed—the logarithmic scaling of perception has proven useful over the last century and a half. For example, dating to the heyday of the Bell Telephone System and still in use today, speech representation for digital telephony has used a logarithmic scaling called the μ-law to match technology to its human end-users. An alternative logarithmic scaling, the A-law, is used in countries other than the United States and Japan. The algorithms used in MP3 compression of music, and in JPEG compression of images, also use logarithmic scaling to appropriately adjust signals for human perception.

But why should the Weber-Fechner and related psychophysical laws exist? Previous explanations have invoked the chemistry of sensory receptors or the informational properties of individual neurons in the brain but did not make connections to the functional requirements of sensory perception. Ernst Weber's original question has until now been unanswered.

in the 1940s and provides a principled framework and mathematical language to study optimal informational systems, whether technological or biological.

One of the central insights from the prehistory of information theory was that any representation of continuous-valued signals must incur some noise or distortion. Perception must obey this rule as well. Thus the ear may receive sound over a range of frequencies, and it transmits that information through communication channels to the brain, where it is processed to form a "sensation." But somewhere along the way, distortion—error—creeps in. The sensation that is the eventual output from the brain must be as robust to—unaffected by—that distortion as possible. In perception, as we have seen, the relative error is the critical one. We want to consider information-processing methodologies that biological information-processing systems might use to optimize expected relative error in the presence of the noise or distortion that necessarily arises.

In formulating a Bayesian model for understanding psychophysical scales, we depend on a branch of information theory called *quantization theory*. All humans quantize on a daily basis; we often round to the nearest dollar at the supermarket or interpret approximate statistics in news articles. Quantization is also an important aspect of fundamental research in science, where physical instruments can only measure a phenomenon to a small number of significant digits, or in engineering for applications where communicating or storing information is costly. In most applications, quantization is thought of as a necessary evil, taking the simple form of dropping less significant digits. However, quantization can be thought of as a more general problem of *compression*: how can I represent the real line with a finite number of points?

We have made the assumption that there is loss of information from where a stimulus is measured at the brain's periphery to when it is perceived at higher cognitive levels because of physical constraints. Our hypothesis is that efficient communication through this channel uses quantization and that the robustness of the system can be studied under a Bayesian framework. Discretizing a continuous signal using a finite number of points leads to stability in information representation as well as robustness to noise and mismatch. We can expect our brain, which is efficient at processing information to display those qualities.

We need a model of the brain in which a few values are communicated well instead of a model where a wide range of values is communicated noisily. It is a quantization model. Our assumptions imply that the brain can actually only distinguish a discrete set of perceptual levels. This notion means that a range of sensory inputs is mapped to a single representative point within the brain. For example, even though there is a continuum of sound intensities that raindrops make when they hit the ground, our brains would only be able to distinguish a finite number of sound levels.

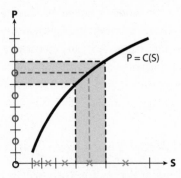

FIGURE 1. Quantization model of psychophysical scales. *S* represents sensory stimulation, such as sound, arriving in a continuous range of intensities. Each intensity range of stimulation is mapped to a discrete point in our perception. Any sound within the shaded range of intensity (between the dashed lines) is perceived as the single circle inside those lines of loudness. The perception points are distributed linearly; the sensation ranges are distributed logarithmically.

We make another assumption: that the quantization at the perceptual level is uniform, meaning that the spacings between elements of perception—of the perceived sound levels in our example—are the same. Figure 1 shows this phenomenon schematically: a continuous range of sensations, *S*, is mapped by the quantizer function to an evenly spaced set of discrete perceptions *P*.

The mathematics of our argument is not complex; Interested readers can find it in our *Journal of Mathematical Psychology* paper (Sun et al. 2013). It turns out that a quantized architecture in the brain that is Bayes optimal for maintaining immunity to relative errors in interpreting signals would map the output—the perceptual space—approximately logarithmically. Hence, we see the logarithmic spacing of the intervals along the horizontal axis in Figure 1. One might expect that the optimal mapping strategy would depend on the statistical distribution of the input, and it does. Remarkably—most remarkably—it does not depend much. Logarithmic perception turns out to be almost optimal no matter what the distribution of intensities of the incoming signal, as long as it is a power law. Remarkably also, it turns out that sensations corresponding to many natural phenomena have statistical distributions that do indeed obey a power law over a range of intensities that are of behavioral interest. George Zipf had first shown that many types of data in the physical and social sciences can be well modeled using this power-law statistical distribution (Zipf 1949). So perception of sound, brightness, pressure, and so on is optimal for relative error if the perception scale is logarithmic.

A Perceptual Model That Uses Signal Compression

The quantization model of perception is simple and intuitive and gives a beautifully concise relationship between the psychophysical scale and stimulus intensity distribution. As we have seen, it provides scales of sensation that are optimal for error. It accounts for why we feel pressures, tastes, loudness, distance, and other sensations on a logarithmic rather than an arithmetic scale. But it may not be suitable for all sensations. For example, numerosity is a perception that may not require sensory inputs that must pass through constrained communication channels. Yet, as we have seen, numbers are still perceived logarithmically. Is this phenomenon still a consequence of optimality as derived through quantization theory? The answer turns out to be yes—as long as the information is further processed and stored in the brain for future use, as numbers often are.

The key idea is that if the input to the quantizer is a continuous random variable, then the output is a discrete random variable. Discrete random variables are easy to compress, and there have been many successful algorithms developed on top of fundamental information theory, for example because of the need to store language texts. Several studies have argued that the brain itself can also use signal compression algorithms. Applying signal compression to the output of the quantizer proposed above yields an amazing result: the optimal choice of perception space is always logarithmic, regardless of the distribution of the stimulus intensity distribution. If compression is used, the optimality of logarithmic scaling laws holds for any stimulus distribution. The statistics of the input do not factor into the psychophysical scale.

A model that includes signal compression seems plausible for perception of numbers but may be less meaningful when connected to external stimuli because properly compressing signals may require long latency and such information can be very time-sensitive.

Returning to the Amazonian Jungle

We have described two models that yield psychophysical scales as Bayes optimal mappings under constraints of communication or of compression and storage. For the first model, power-law distributions of stimuli do indeed yield the Weber-Fechner logarithmic law, but we might wonder what happens when we look at actual, empirical, stimulus data. After all, data yields true tests of theoretical predictions.

One well-studied sensation is the loudness of sound. Loudness is one auditory cue that conveys information related to mating rituals, predator warnings, and locations for food. It is likely that auditory systems have evolved to process natural sounds efficiently because it is essential for survival and reproduction. Experiments have shown that the human ear can perceive loudness at differences of about 1 dB, with small deviations at extreme ranges; this is consistent with the Weber-Fechner law.

To test our theory of optimal psychophysical scales, we return to the Amazonian rain forest of the Mundurucu: we tested our model on data sets of animal vocalization and human speech. The animal data set features recordings of rain forest mammals from commercially available CDs; the human data set is of a male speaker reciting English sentences. In both data sets, we removed silences and then extracted sound intensity over small time windows. We then estimated the distribution of sound intensities using the empirical data. Feeding this empirical distribution of sound intensities into our quantization model, we can predict the best psychophysical scale for minimizing expected relative error.

Our results are shown in Figures 2 and 3. They demonstrate that the psychophysical scale—the perceived loudness—is approximately linear when the stimulus intensity is plotted on a log scale, which does indeed indicate that, when the expected relative error in a signal is minimized, there is an approximate logarithmic relationship between the stimulus and perception.

Conclusion

The field of psychophysics considers questions of human sensory perception, and one of its most robust findings has been the Weber-Fechner law. It has helped us understand *how* we perceive the world. In putting

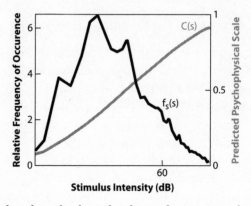

FIGURE 2. Predicted psychophysical scale $C(s)$ from empirical intensity distribution $f_s(s)$ of rain forest sounds.

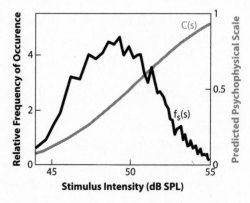

FIGURE 3. Predicted psychophysical scale $C(s)$ from empirical intensity distribution $f_s(s)$ of human speech.

forth an optimality theory for understanding *why* we perceive things as we do, we have come to a few general ideas.

Our work points out the importance of the statistics of the stimulus in interpreting psychophysical laws. Indeed, our theory allows one to predict the logarithmic scales of our perceptions simply from the statistics of the natural world. Just as Gustav Fechner had suggested in the early days, internal representations are fundamentally intertwined with the external world they describe. With an optimality result, one can say that people and animals are good processors of statistical signals—or at least our sensing apparatus seems to have evolved to be so.

Broadly speaking, our findings support the possibility that animals have information-theoretically optimal signal acquisition and perception structures at a cognitive level. Going forward, it is important to make further experimental tests of our optimality hypothesis. For example, there are certain specialized sensory modes that are known experimentally not to have logarithmic psychophysical scales. Night vision is one. Optimizing for the corresponding stimulus distributions would provide a way to verify or falsify our optimal information-processing hypothesis.

References

Dehaene, Stanislas. (2011). *The Number Sense: How the Mind Creates Mathematics*, New York: Oxford University Press.

Fechner, Gustav Theodor. (1860). *Elemente der Psychophysik*, Leipzig, Germany: Breitkopf & Härtel.

Sun, John Z., Wang, Grace I., Goyal, Vivek K., and Varshney, Lav R. (2012). "A framework for Bayesian optimality of psychophysical laws." *Journal of Mathematical Psychology*, **56**(6), 495–501.

Varshney, Lav R., Sjostrom, P. J., and Chklovskii, D. B. (2006). "Optimal information storage in noisy synapses under resource constraints." *Neuron*, **52**(3), 409–23.

Zipf, George K. (1949). *Human Behavior and the Principle of Least Effort*, Cambridge, MA: Addison-Wesley.

The Music of Math Games

KEITH DEVLIN

Search online for video games and apps that claim to help your children (or yourself) learn mathematics, and you will be presented with an impressively large inventory of hundreds of titles. Yet hardly any survive an initial filtering based on seven very basic pedagogic "no-nos" that any game developer should follow if the goal is to use what is potentially an extremely powerful educational medium to help people learn math. A good math learning game or app should avoid:

- Confusing mathematics itself (which is really a way of thinking) with its representation (usually in symbols) on a flat, static surface.
- Presenting the mathematical activities as separate from the game action and game mechanics.
- Relegating the mathematics to a secondary activity, when it should be the main focus.
- Adding to the common perception that math is an obstacle that gets in the way of doing more enjoyable activities.
- Reinforcing the perception that math is built on arbitrary facts, rules, and tricks that have no unified, underlying logic that makes sense.
- Encouraging students to try to answer quickly, without reflection.
- Contributing to the misunderstanding that math is so intrinsically uninteresting that it has to be sugar-coated.

Of the relatively few products that pass through this seven-grained filter—which means they probably at least don't do too much harm—the majority focus not on learning and understanding but on mastering basic skills, such as the multiplicative number bonds (or "multiplication tables"). Such games don't actually provide learning at all, but they do make good use of video game technology to take out of the classroom the acquisition of rote knowledge. This use leaves the teacher more time and freedom to focus on the main goal of mathematics teaching, namely, the development of what I prefer to call "mathematical thinking." Many people have come to believe mathematics is the memorization of, and mastery at using, various formulas and symbolic procedures to solve encapsulated and essentially artificial problems. Such people typically have that impression of math because they have never been shown anything else. If mention of the word *algebra* automatically conjures up memorizing the use of the formula for solving a quadratic equation, chances are you had this kind of deficient math education. For one thing, that's not algebra but arithmetic; for another, it's not at all representative of what algebra is, namely, thinking and reasoning about entire classes of numbers, using logic rather than arithmetic.

What's in a Game?

So how to go about designing a good video game to help students learn mathematics? The first step should be to read—several times, from cover to cover—the current "bible" on K-12 mathematics education. It is called *Adding It Up: Helping Children Learn Mathematics,* and it was published by the National Academies Press in 2001. The result of several years' work by the National Research Council's Mathematics Learning Study Committee, a blue-ribbon panel of experts assembled to carry out that crucial millennial task, this invaluable volume sets out to codify the mathematical knowledge and skills that are thought to be important in today's society. As such, it provides the best single source currently available for guidelines on good mathematics instruction.

The report's authors use the phrase *mathematical proficiency* to refer to the aggregate of mathematical knowledge, skills, developed abilities, habits of mind, and attitudes that are essential ingredients for life in the 21st century. They break this aggregate down to what they describe as "five tightly interwoven" threads. The first is *conceptual understanding,* the comprehension of mathematical concepts, operations, and relations. The second is *procedural fluency,* defined as skill in carrying out arithmetical procedures accurately, efficiently, flexibly, and appropriately. Third is *strategic competence,* or the ability to formulate, represent, and solve mathematical problems arising in real-world situations. Fourth is *adaptive reasoning*—the capacity for logical thought, reflection, explanation, and justification. Finally there's *productive disposition,* a habitual inclination to see mathematics as sensible, useful, and worthwhile, combined with a confidence in one's own ability to master the material.

The authors stress that it is important not to view these five goals as a checklist to be dealt with one by one. Rather, they are different aspects of what should be an integrated whole, with all stages of teaching focused on all five goals.

So it's not that the crucial information about mathematics learning required to design good learning video games is not available—in a single, eminently readable source—it's that few people outside the math education community have read it.

Combining Skills

The majority of video games designed to provide mathematics learning fail educationally for one of two reasons: Either their designers know how to design and create video games but know little about mathematics education (in particular, how people learn mathematics) and in many cases don't seem to know what math really is, or they have a reasonable sense of mathematics and have some familiarity with the basic principles of mathematics education but do not have sufficient experience in video game design. (Actually, the majority of math education games seem to have been created by individuals who know little more than how to code, so those games fail both educationally and as games.)

To build a successful video game requires an understanding, at a deep level, of what constitutes a game, how and why people play games, what keeps them engaged, and how they interact with the different platforms on which the game is to be played. That is a lot of deep knowledge.

To build an engaging game that also supports good mathematics learning requires a whole lot more: understanding, at a deep level, what mathematics is, how and why people learn and do mathematics, how to get and keep them engaged in their learning, and how to represent the mathematics on the platform on which the game is to be played. That, too, is a lot of deep knowledge.

FIGURE 1. In the author's game, *Wuzzit Trouble,* the cute and fuzzy creatures must be freed from traps controlled by gearlike combination locks. Players collect keys to open the locks by solving puzzles of varying difficulty. See also color image. (Image courtesy of BrainQuake Inc.)

In other words, designing and building a good mathematics educational video game—be it a massively multiplayer online (MMO) game or a single smartphone app—requires a team of experts from several different disciplines. That means it takes a lot of time and a substantial budget. How much? For a simple-looking, casual game that runs on an iPad, reckon nine months from start to finish and a budget of $300,000.

Following the tradition of textbook publishing, that budget figure does not include any payment to the authors, who essentially create the entire pedagogic framework and content, nor the project's academic advisory board (which it should definitely have).

The Symbol Barrier

Given the effort and the expense to make a math game work, is it worth the effort? From an educational perspective, you bet it is. Though the vast majority of math video games on the market essentially capitalize on just one educationally important aspect of video games—their power to fully engage players in a single activity for long periods of time—all but a tiny number of games (fewer than 10 by my count) take advantage of another educationally powerful feature of the medium: video games' ability to overcome the *symbol barrier*.

Though the name is mine, the symbol barrier has been well known in math education circles for more than 20 years and is recognized as the biggest obstacle to practical mastery of middle school math. To understand the symbol barrier and appreciate how pervasive it is, you have to question the role symbolic expressions play in mathematics.

By and large, the public identifies doing math with writing symbols, often obscure symbols. Why do they make that automatic identification? A large part of the explanation is that much of the time they spent in the mathematics classroom was devoted to the development of correct symbolic manipulation skills, and symbol-filled books are the standard way to store and distribute mathematical knowledge. So we have gotten used to the fact that mathematics is presented to us by way of symbolic expressions.

But just how essential are those symbols? After all, until the invention of various kinds of recording devices, symbolic musical notation was the only way to store and distribute music, yet no one ever confuses music with a musical score.

Just as music is created and enjoyed within the mind, so too is mathematics created and carried out (and by many of us enjoyed) in the mind. At its heart, mathematics is a mental activity—a way of thinking—one that over several millennia of human history has proved to be highly beneficial to life and society.

In both music and mathematics, the symbols are merely static representations on a flat surface of dynamic mental processes. Just as the trained musician can look at a musical score and hear the music come alive in her or his head, so too the trained mathematician can look at a page of symbolic mathematics and have that mathematics come alive in the mind.

So why is it that many people believe mathematics itself is symbolic manipulation? And if the answer is that it results from our classroom experiences, why is mathematics taught that way? I can answer that second question. We teach mathematics symbolically because, for many centuries, symbolic representation has been the most effective way to record mathematics and pass on mathematical knowledge to others.

Still, given the comparison with music, can't we somehow manage to break free of that historical legacy?

Though the advanced mathematics used by scientists and engineers is intrinsically symbolic, the kind of math important to ordinary people in their lives—which I call everyday mathematics—is not, and it can be done in your head. Roughly speaking, everyday mathematics comprises counting, arithmetic, proportional reasoning, numerical estimation, elementary geometry and trigonometry, elementary algebra, basic probability and statistics, logical thinking, algorithm use, problem formation (modeling), problem solving, and sound calculator use. (Yes, even elementary algebra belongs in that list. The symbols are not essential.)

True, people sometimes scribble symbols when they do everyday math in a real-life context. But for the most part, what they write down are the facts needed to start with, perhaps the intermediate results along the way, and, if they get far enough, the final answer at the end. But the doing-math part is primarily a thinking process—something that takes place mostly in your head. Even when people are asked to "show all their work," the collection of symbolic expressions that they write down is not necessarily the same as the process that goes on in their minds when they do math correctly. In fact, people can become highly skilled at doing mental math and yet be hopeless at its symbolic representations.

With everyday mathematics, the symbol barrier emerges. In their 1993 book *Street Mathematics and School Mathematics,* Terezinha Nunes, David William Carraher, and Analucia Dias Schliemann describe research carried out in the street markets of Recife, Brazil, in the early 1990s. This and other studies have shown that when people are regularly faced with everyday mathematics in their daily lives, they rapidly master it to an astonishing 98 percent accuracy. Yet when faced with what are (from a mathematical perspective) the very same problems, but presented in the traditional symbols, their performance drops to a mere 35 to 40 percent accuracy.

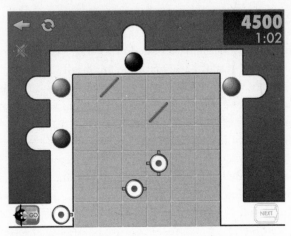

FIGURE 2. *KickBox* uses a penguin character called JiJi that players must help get from one end of the corridor to the other. Players position beam-splitters and reflectors to direct lasers that knock out obstacles in JiJi's path. Solving such a puzzle provides excellent practice in mathematical thinking, completely separate from the more familiar formulas, equations, and dreaded "word problems." See also color image. (Screenshot image courtesy of MIND Research Institute.)

It simply is not the case that ordinary people cannot do everyday math. Rather, they cannot do symbolic everyday math. In fact, for most people, it's not accurate to say that the problems they are presented in paper-and-pencil format are "the same as" the ones they solve fluently in a real-life setting. When you read the transcripts of the ways they solve the problems in the two settings, you realize that they are doing completely different things. Only someone who has mastery of symbolic mathematics can recognize the problems encountered in the two contexts as being "the same."

The symbol barrier is huge and pervasive. For the entire history of organized mathematics instruction, where we had no alternative to using static, symbolic expressions on flat surfaces to store and distribute mathematical knowledge, that barrier has prevented millions of people from becoming proficient in a cognitive skill set of evident major importance in today's world, on a par with the ability to read and write.

FIGURE 3. *Motion Math* is a *Tetris*-inspired game that uses the motion sensors in a smartphone or tablet to allow players to tilt the screen to direct descending fractions to land on the right location on the number line. This game is an excellent introduction to fractions for younger children because it connects the abstract concept to tactile, bodily activity. See also color image. (Image courtesy of Motion Math.)

Going Beyond

With video games, we can circumvent the barrier. Because video games are dynamic, interactive, and controlled by the user yet designed by the developer, they are the perfect medium for representing everyday mathematics, allowing direct access to the mathematics (bypassing the symbols) in the same direct way that a piano provides direct access to the music.

It's essentially an interface issue. Music notation provides a useful interface to music, but it takes a lot of learning to be able to use it. It's the same for mathematics notation.

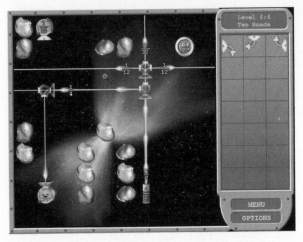

FIGURE 4. In the math puzzle game *Refraction,* players learn about fractions and algebra. In this puzzle, the player has to split a laser beam a sufficient number of times to power all of the alien spaceships on the screen. The game is also designed to be modified on the fly, in an effort to capture data about what teaching methods and reward systems work best for students. See also color image. (Image courtesy of the Center for Game Science, University of Washington.)

The piano provides an interface to music that is native to the music, and hence far more easy and natural to use. When properly designed, video games can provide interfaces to mathematical concepts that are native to those concepts, and thus far more easy and natural to use.

Consider some of the reasons so many people are able to master the piano. You learn by doing the real thing (initially poorly, on simple tunes, but getting better over time). You use the very same instrument on Day 1 that the professionals use. You get a sense of direct involvement with the music. You get instant feedback on your performance— the piano tells you if you are wrong and how you are wrong, so you can gauge your own progress. The instructor is your guide, not an arbitrator of right or wrong. And the piano provides true *adaptive learning.*

We read a lot today about adaptive learning, as if it were some new invention made possible by digital technologies. In fact, it is a proven method that goes back to the beginning of human learning.

What's more, the proponents of today's digital version have gotten it all wrong, and as a result produce grossly inferior products. They try to use artificial intelligence so that an "educational delivery system" can modify the delivery based on the student's performance.

Yet tens of thousands of years of evolution have produced the most adaptive device on the planet: the human brain. Trying to design a computer system to adapt to a human's cognitive activity is like trying to build a cart that will draw a horse. Yes, it can be done, but it won't work nearly as well as building a cart that a horse can pull.

The piano metaphor can be pursued further. There's a widespread belief that you first have to master the basic skills to progress in mathematics. That's total nonsense. It's like saying you have to master musical notation and the performance of musical scales before you can start to try to play an instrument—a surefire way to put someone off music if ever there was one. Learning to play a musical instrument is much more enjoyable, and progress is much faster, if you pick up—and practice—the basic skills as you go along, as and when they become relevant and important to you. Likewise, for learning mathematics, it's not that basic skills do not have to be mastered, but rather it's how the student acquires that mastery that makes the difference.

When a student learning to play the piano is faced with a piece she or he cannot handle, the student (usually of his or her own volition) goes back and practices some easier pieces before coming back to the harder one. Or perhaps the learner breaks the harder piece into bits and works on each part, at first more slowly, then working up to the correct tempo. What the player does not do is go back to a simpler piano (one with fewer keys, perhaps?); nor do we design pianos that somehow become easier to play. The piano remains the same; the players adjust (or adapt) what they do at each stage. The instrument's design allows use by anyone, from a rank beginner to a concert virtuoso.

This lesson is the one we need to learn to design video games to facilitate good mathematics learning. For more than 2,000 years, commentators have observed connections between mathematics and music. We should extend the link to music when it comes to designing video games to help students learn math, thinking of a video game as an instrument on which a person can "play" mathematics.

A Mathematical Orchestra

The one difference between music and math is that whereas a single piano can be used to play almost any tune, a video game designed to play, say, addition of fractions, probably won't be able to play multiplication of fractions. This means that the task facing the game designer is not to design one instrument but an entire orchestra.

Can this be done? Yes. I know this fact to be true because I spent almost five years working with talented and experienced game developers on a stealth project at a large video game company, trying to build such an orchestra. That particular project was eventually canceled but not because we had not made progress—we had developed more than 20 such "instruments"—but because the pace and cost of development did not fit the company's entertainment-based financial model. A small number of us from that project took all that we had learned and formed our own company, starting from scratch to build our own orchestra.

In the meantime, a few other companies have produced games that follow the same general design principles we do. Some examples include the games *Motion Math* and *Motion Math Zoom,* which use the motion sensors in a smartphone or tablet to allow players to interact directly with numbers. The puzzle game *Refraction* was produced by a group of professors and students in the Center for Game Science at the University of Washington and was designed as a test platform that could be altered on the fly to see what teaching methods and reward systems work best for students learning topics such as fractions and algebra. *DragonBox* focuses on learning algebra in a puzzle where a dragon in a box has to be isolated on one side of the screen. *KickBox* uses physical concepts—such as positioning lasers to get rid of obstacles for the game's penguin mascot—to learn math concepts. The same producer, the MIND Research Institute, also developed *Big Seed,* a game where players have to unfold colored tiles to completely fill a space. These games all combine the elements of math learning with game play in an effective, productive fashion.

The game produced by my colleagues and me, *Wuzzit Trouble* is a game where players must free the Wuzzits from the traps they've inadvertently wandered into inside a castle. Players must use puzzle-solving skills to gather keys that open the gearlike combination locks on the cages, while avoiding hazards. As additional rewards, players can give the Wuzzits treats and collect special items to show in a "trophy room."

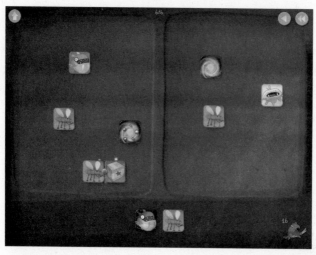

FIGURE 5. *DragonBox* challenges players to isolate the glittering box (containing a growling dragon) on one side of the screen. What they are doing is solving for the *x* in an algebraic equation. But there isn't an *x* to be seen in the early stages of the game. As the player progresses through the game, mathematical symbols start to appear, first as names for the objects, later replacing the object altogether. This game demonstrates very clearly that solving an algebraic equation is not fundamentally about manipulating abstract symbols but is rather reasoning about things in the world, for which the symbols are just names. *DragonBox* provides a more user-friendly interface to algebraic equations—but it's still algebra, and even young children can do it. See also color image. (Image courtesy of WeWantToKnow AS.)

We worked with experienced game developers to design *Wuzzit Trouble* as a game that people will want to play purely for fun, though admittedly mentally challenging, puzzle entertainment. So it looks and plays like any other good video game you can play on a smartphone or tablet. But unlike the majority of other casual games, it is built on top of sound mathematical principles, which means that anyone who plays it will be learning and practicing good mathematical thinking—much like a person playing a musical instrument for pleasure will at the same time learn about music. Our intention is to provide, separately and at a later date, suggestions to teachers and parents for how to use the game as a basis for more formal learning. *Wuzzit Trouble* might look and play like a simple arithmetic game, and indeed that is the point. But looks can be deceiving. The puzzles carry star ratings, and I have yet to achieve the maximum number of stars on some of the puzzles! (I never mastered Rachmaninoff on the piano either.) The game is not designed to teach. The intention is to provide an "instrument" that, in addition to being fun to play, not only provides implicit learning but may also be used as a basis for formal learning in a scholastic setting. We learned all of these design lessons from the piano.

The Fundamental Theorem
of Algebra for Artists

BAHMAN KALANTARI AND BRUCE TORRENCE

It's simple. It's beautiful. And it doesn't get more fundamental: Every polynomial factors completely over the complex numbers. So says the fundamental theorem of algebra.

If you are like us, you've been factoring polynomials more or less since puberty. What may be less clear is why. The short answer is that polynomials are the most basic and ubiquitous functions in existence. They are used to model all manner of natural phenomena, so solving polynomial equations is a fundamental skill for understanding our world. And thanks to the fundamental theorem, solving polynomial equations comes down to factoring. However, finding the factors is often easier said than done.

For instance, start with your favorite polynomial equation, something like $x^3 - 2x^2 - 5x + 5 = -1$. Any such equation can be rewritten so that it is equal to zero; in this case, we add one to each side: $x^3 - 2x^2 - 5x + 6 = 0$. The fundamental theorem says the polynomial on the left factors completely. With a bit of work we get $(x - 1)(x + 2)$ $(x - 3) = 0$. Since a product of numbers is zero if and only if one of those numbers is itself zero, the factored form tells us the solutions to the original equation: $x = 1$, $x = -2$, and $x = 3$.

But wait. What about $x^2 + 1$? It can't be factored, you may protest, since a square plus one cannot be zero. Although this is true for *real* numbers x, this polynomial (and according to the fundamental theorem, *all* polynomials) does factor over the complex numbers. In other words, to factor $x^2 + 1$, mathematicians first had to ask: Is there a larger domain of numbers where it could be factored? And after a long time— several centuries into the enterprise of solving polynomial equations— the complex numbers were discovered. These are numbers of the form $a + ib$, where a and b are real numbers and i is an "imaginary" number satisfying $i^2 = -1$. Complex numbers can be thought of as points in the Euclidean plane by associating the number $a + ib$ with the ordered pair (a, b). So complex numbers turn points in the plane into numbers to which we can apply the four elementary operations of addition, subtraction, multiplication, and division. This was a profound discovery. In this case, we have $x^2 + 1 = (x + i)(x - i)$.

We are leaving the comforts of the real number system behind, so let's agree to use z rather than x for the variable name. It's a time-honored notational convention that z represents a complex variable.

Whether over the real or complex numbers, factoring is hard. Given n complex numbers $z_1, z_2, \ldots, z_n$ (think of them as points in the complex plane), it is a cinch to construct a polynomial with those roots: Just expand the product $(z - z_1)(z - z_2) \cdots (z - z_n)$. But suppose we are given a polynomial of degree n, that is, a function of the form

$$p(z) = a_n z^n + a_{n-1} z^{n-1} + \cdots + a_1 z + a_0$$

where the leading coefficient $a_n \neq 0$ and the coefficients a_i may be real or complex. Finding its roots, i.e., factoring it—finding those n points in the plane—can be a formidable challenge.

Mathematicians and computer scientists have developed sophisticated algorithms to find, or at least approximate, those pesky roots. The fundamental theorem is an important ingredient in those algorithms. It guarantees that a search for the roots is not in vain.

Visualizing a Complex Polynomial

It's easy to graph a function f whose domain and range are both real. Use the x axis for the domain, the y axis for the range. The graph consists of all points of the form $(x, f(x))$. Easy.

It's not so easy to graph a function when the domain and range are both complex. The domain and range are each two-dimensional, so four dimensions are required to carry out the analog of the traditional graph. Bummer.

Some compromises facilitate visualization. Here's one that's based on *modulus*. The modulus $|z|$ of a complex number $z = x + iy$ is its distance from the origin, namely $\sqrt{x^2 + y^2}$. It is the complex analog of absolute value. To make a modulus graph of a complex function

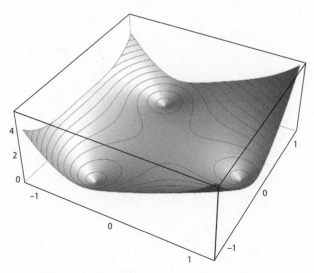

FIGURE 1. The 3-D modulus graph of $p(z) = z^3 - 1$. See also color image.

p, use the x-y plane to represent the domain, so that the point (x, y) corresponds to the complex number $x + iy$. Use the vertical axis to represent the modulus of the complex number $p(x + iy)$. Since distance cannot be negative, this 3-D graph never dips below the x-y plane. For example, Figure 1 shows the graph we get for the complex polynomial $p(z) = z^3 - 1$. Writing $z = x + iy$, you can check that the modulus of $z^3 - 1$ is

$$\sqrt{(x^3 - 3xy^2 - 1)^2 + (3x^2y - y^3)^2}.$$

The three dimples dip down to touch the x–y plane and represent the roots of this polynomial: the three cube roots of unity. The fundamental theorem implies that the modulus graph of any polynomial must dip down and touch the x-y plane.

Given the two-dimensional nature of the page you are looking at, we can improve the readability of a modulus graph by using a 2-D contour plot instead. Figure 2 shows the previous graph as a contour plot, with a white dot at each root.

In addition to the three roots, there is another interesting point on this graph: $z = 0$ (the origin). When $z = 0$, note that $p(0) = 0^3 - 1 = -1$,

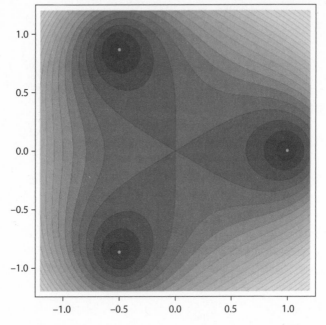

FIGURE 2. A contour plot of the modulus of $p(z) = z^3 - 1$. See also color image.

which has modulus one. Look at the level curve representing all points of modulus one. If you imagine a small disk centered at the origin on the plot (as in Figure 3), the level curve through the origin divides the disk into six equal sectors.

Judging by the shadings of the contour regions, three of those regions head down, toward the three roots, whereas the other three head up toward higher ground. Moreover, the sectors alternate between ascent and descent. This is a manifestation of a beautiful property of polynomials called the *geometric modulus principle*.

The Geometric Modulus Principle

Before giving an explicit statement of the principle, we'll whet your appetite by explaining how one could predict that for the function $p(z) = z^3 - 1$, a disk centered at the origin would be divided into six ascent/descent sectors. Why six? We first take a few derivatives and evaluate

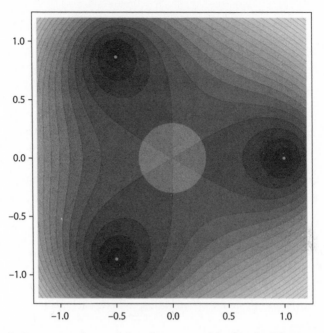

FIGURE 3. Examining the modulus plot in a neighborhood of the origin. See also color image.

them at the point in question: $z = 0$. Mercifully, differentiation in the complex landscape works just as in first-semester calculus:

$$p'(z) = 3z^2,$$
$$p''(z) = 6z,$$
$$p^{(3)}(z) = 6.$$

Note that $z = 0$ is a root of p' and p''. The *third* derivative is the first that does not vanish when we plug in $z = 0$. For this example, the geometric modulus principle says that each ascent/descent sector of the disk centered at $z = 0$ has central angle $\pi/3$, where the three appears in the denominator because the *third* derivative is the first to not vanish. A central angle of $\pi/3$ produces six sectors.

More generally, suppose we are given any complex polynomial $p(z)$ of positive degree n and a complex number z_0. Imagine a small disk centered at z_0 in the modulus plot of $p(z)$. If z_0 is a root of p, then *every* direction from z_0 is a direction of ascent; move a small amount in any

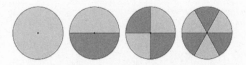

FIGURE 4. Disks shown at a root (left) and at points with $k = 1$, $k = 2$, and $k = 3$.

direction from z_0, and the modulus of p changes from 0 to a positive value. On the other hand, if z_0 is not a root of p, consider the derivatives

$$p'(z_0), p''(z_0), p^{(3)}(z_0), \ldots$$

and let k be the position of the first nonzero number in this list. Note that since p is a polynomial of degree n, the nth derivative must be a nonzero constant, so $1 \leq k \leq n$. The geometric modulus principle says that a small disk centered at z_0 is divided into equal alternating sectors of ascending and descending modulus, where each sector has central angle π/k.

Looking again at the modulus plot of $z^3 - 1$, if one chooses *any* point z_0 other than 0 or one of the three roots, then $p'(z_0) \neq 0$, and so $k = 1$. A small disk centered at z_0 is split precisely in half. The split occurs along the tangent line to the level curve through z_0; there is nothing particularly special here.

The interesting part concerns those points where $k > 1$, the *critical points* of p. That disks centered at these points should be *evenly* divided into sectors of ascending and descending modulus is both surprising and beautiful (Figure 4). A quick survey of the functions one encounters in a multivariable calculus course—real-valued functions of two real variables—reveals that typical saddle points do not have equal sectors of ascent and descent. The modulus functions for complex polynomials are quite special in this sense.

The Fundamental Theorem

There are numerous proofs of the fundamental theorem of algebra, and several make use of the modulus function for a polynomial $p(z)$. The basic idea behind such proofs is to first note that when z is sufficiently far from the origin, $|p(z)|$ must be large. On the other hand, within the confines of a large closed disk centered at the origin, $|p(z)|$ must attain both a minimum and a maximum. Because the values get large as one approaches the edge of the disk, all minima must occur in its interior, and so there must be a global minimum.

The proofs then argue that for any point z_0 that is not a root of p, there is a direction of descent for the modulus function, and so that point cannot be a minimum. There is then only one possibility for a minimum of the modulus function: It must be a root of p. The last step is to note that once a single root of p has been found, say z_1, it can be factored out: $p(z) = (z - z_1)q(z)$ for some polynomial $q(z)$. But q then has a root by the same reasoning, and in a few quick steps we have factored p completely.

This line of proof can be harnessed to easily prove the fundamental theorem of algebra once the geometric modulus principle is established. In other words, the fundamental theorem is a simple logical consequence of the geometric modulus principle. Other deep results can also be derived as a consequence, such as the Gauss-Lucas theorem and the maximum modulus principle for polynomials. It is not a stretch to characterize the geometric modulus principle as something fundamental in its own right.

Fundamental Art

The geometric modulus principle provides a striking visual facet to the inherently abstract realm of complex polynomials. Given any collection of n complex numbers, make a modulus plot for a polynomial with roots at those numbers. If level curves—curves of constant modulus— passing through the critical points are added, the geometric modulus principle comes to life.

In Figures 5 and 6, roots are shown as small, light dots, and critical points (that are not also roots) are black dots. Each image is the modulus plot for a polynomial, showing only those level curves that pass through critical points. The geometric modulus principle guarantees that small disks centered at the critical points are always divided equally into sectors by those level curves. And of course, level curves representing distinct moduli can never touch one another.

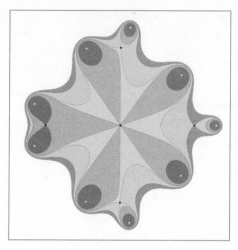

FIGURE 5. Modulus plot of $p(z) = z^9 - z^5 - 1$. See also color image.

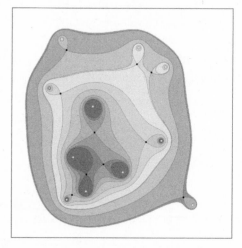

FIGURE 6. Modulus plot of a random polynomial of degree 11. See also color image.

You can also take a collection of points in the complex plane and center a disk at each point. For each disk, divide it into some even number of sectors of equal size (rotated any way you like), and alternately label each sector "ascent" or "descent" (Figure 7). There is guaranteed to be a polynomial whose modulus plot matches all

FIGURE 7. There is a polynomial whose modulus plot conforms to the ascent and descent sectors of these disks!

the ascent/descent sectors at the indicated points! It may have additional roots and critical points, but it's guaranteed to fit your choices precisely.

Further Reading

The geometric modulus principle was introduced and proved in the first author's paper, "A geometric modulus principle for polynomials," which appeared in the *American Mathematical Monthly* 118 (2011); 931–35.

The Arts—Digitized, Quantified, and Analyzed

NICOLE LAZAR

The inspiration for this column is an article I came across on the website of *Financial Times* from June 15, 2013, titled "Big Data Meets the Bard."The focus of the article is Stanford University's "Literary Lab" run by Franco Moretti. In the lab, Moretti, his colleagues, and his students use data analysis techniques—and especially tools that we have come to associate with Big Data—to analyze plays, books, authors, and entire genres of literature. This got me thinking more broadly about the intersections between statistics and the arts, and between Big Data and the arts, which here I'll define to be literature, music, and the visual arts (e.g., sculpture, painting).

Of course, the use of statistics for analysis of literary texts is not new. Already in 1963, Fred Mosteller and David Wallace published their now-famous study of the disputed Federalist papers, using (Bayesian) statistical methods to perform inference on authorship attribution. Much like the work done in the Moretti lab for genre attribution, Mosteller and Wallace looked at word counts as a key feature. Likewise, Bradley Efron and Ronald Thisted used statistical methods to estimate the number of words Shakespeare knew, as well as to study authorship questions about poems that may or may not have been written by the Bard. The scope of these projects, however, was much smaller and did not rely on text mining of the type that is possible with modern computers.

Much more recently, Big Data methods were used to show that J. K. Rowling, of *Harry Potter* fame, was the most likely true identity of Robert Galbraith, author of the best-selling murder mystery novel *The Cuckoo's Calling*. The features used in this discovery were the distribution of word length, the 100 most commonly used words, distribution of character 4-grams (that is, strings of four characters that appear consecutively in or across words, such as "ring" or "ofth") and distribution of word bigrams (pairs of words that appear together). Details on the analysis can be found at languagelog.ldc.upenn.edu/nll/?p=5315.

Beyond literature and text analysis, statistics has been brought to bear on the analysis of musical pieces, and the twentieth-century composer Iannis Xenakis (1922–2001) used probability theory and stochastic processes in composing his music. Statistical techniques also have been used, although more rarely, in the analysis of paintings to attribute new or disputed works to particular artists—for example, to distinguish forgeries from authentic paintings by Vincent Van Gogh.

Still, the arts are not the first application that comes to mind when one mentions statistics! Nor are the connections between art and Big Data immediately apparent. But they are there. I've come up with three perspectives on these connections. First is the straightforward use of Big Data techniques to analyze literature and art, as I have discussed in the opening paragraphs. Second is the creation of art from data, big or small—this use expands on existing ideas of data visualization and takes them to a higher level, beyond just conveying statistical information. Third is the analysis of data about art, what might be crassly termed "arts analytics" in analogy to business analytics—analysis of record sales, artist popularity, and so forth.

My starting point in this journey, then, was the work of the Literary Lab. The article in the *Financial Times* piqued my curiosity, and so I visited their website at litlab.stanford.edu. The group has published "pamphlets" (what we would probably call "technical reports" in statistics departments), and all are available for download. A summary of their contents showcases the variety of projects in which members of the group are involved.

Pamphlet Number 1, titled "Quantitative Formalism: An Experiment," is, in some ways, proof of concept of the statistical approach to text analysis. The team used unsupervised learning techniques—cluster analysis—to group Shakespeare's plays and nineteenth-century British novels by genre. Intriguingly, classifications that agree quite strongly with the opinions of experts on how certain texts should be classified (for example, that *Macbeth* is a tragedy, or that *Ivanhoe* is a historical novel) can be obtained via cluster analysis on purely lexical features—e.g., the most frequently used words and punctuation marks.

That is, the use of a word such as "the" or "he" is different enough in a gothic novel than in an industrial novel that cluster analysis can tell the genres apart. Similarly, Shakespeare's comedies are clustered separately from tragedies, for instance, based on the most frequent words (although *Othello* oddly enough is clustered with the comedies and *Hamlet* with the histories!). Apparently, the differences between genres are not solely evident at the macroscopic levels of themes and keywords that we, as readers, might use, but also in the micro level of seeming junk words such as "and" and "but."

Pamphlet Number 2, "Network Theory, Plot Analysis" examines *Hamlet* through the lens of social network analysis. Every character in the play is a node, and there is an edge between two characters if any words are exchanged between them over the course of the play. It is worth noting that other ways of defining relationships between nodes are possible. Some interesting patterns are revealed: between them, Hamlet and Claudius interact directly with almost all the characters; however, the same is not true for Ophelia and Gertrude, reflecting the smaller social frame of the women in the play. Almost all the characters who interact directly with both Hamlet and Claudius are killed, but nobody else dies. There is a so-called "zone of death" around the warring king and prince. Although Hamlet, Claudius, and Horatio are all strong nodes in the network, having the most connections to other players, the implications of removing each are vastly different. Exploring the play from the perspective of social networks reveals some interesting relationships and structure. Even my husband—who has a Ph.D. in English—was intrigued by the "zone of death" uncovered by this analysis.

The third pamphlet, "Becoming Yourself: The Afterlife of Reception," studies the work of the late David Foster Wallace, a young novelist and essayist who committed suicide in 2008. Here, too, the approach is a network analysis, this time based on Amazon recommendations and reviews in the mainstream press. The goal is to explore comparisons, distinctions, and connections made between Wallace and his contemporaries, such as Michael Chabon (*Wonder Boys, The Mysteries of Pittsburgh, The Amazing Adventures of Kavalier & Clay*) and Jonathan Franzen (*The Corrections, How to Be Alone*) and stylistic brethren such as Thomas Pynchon (*Gravity's Rainbow, The Crying of Lot 49*). The main findings of the network analysis are that (a) Wallace is different from other writers of his generation in that his books cluster—readers of Wallace read more Wallace; (b) Wallace's style is decidedly postmodern, reflecting the work of Pynchon, DeLillo, etc., rather than authors of his own generation; and (c) Wallace is integral in that he unites readers with diverse interests and backgrounds.

Pamphlet Number 4, "A Quantitative Literary History of 2,958 Nineteenth-Century British Novels: The Semantic Cohort Method" studies how social shifts in Britain are reflected in the language and style of novels written during this period. The work develops quantitative methods for "tracing changes in literary language." The authors of this report describe their effort as a high-level study of cultural history, using computational methods: How does language use change over time in written texts, and what does this tell us about concomitant changes in the culture? They use the example of Google's Ngram viewer, which shows the frequency of word usage in written texts and is a popular tool for studying just those changes. The analysis approach in this report is based on building semantic fields—collections of words that can be studied over time to indicate changes in the themes (in this case) of novels, essentially the extraction of features. Some of the generated fields include "social restraint" (words relating to the value of moderation in social behavior) and "sentiment" (words relating to emotions). One finding is that the use of words from the social restraint field decreased roughly linearly from 1785 to 1900 (in the novels in the examined corpus), presumably reflecting societal trends in the perceived importance of this characteristic of behavior.

Pamphlet Number 5, "Style at the Scale of the Sentence," examines sentence types at the microscopic level—what sorts of sentences do different authors use, for one, but also what sorts of verbs and words more generally are associated with different sentence types? Plots of these look almost like images you may have seen of DNA sequences. Based on these building blocks, the "semantic space" of sentences is mapped using the first two principal components, leading eventually to a mapping by genre. The results presented in this report are complementary to those in the first pamphlet and build on them. The moods we associate with a particular genre—the broodiness and fear of a gothic novel, for instance—come from word choices, but also from how those words are used. Verb tenses are different in a romance of manners than in a gothic romance, perhaps not an Earth-shattering

PAMPHLETS

5. STYLE AT THE SCALE OF THE SENTENCE
[DOWNLOAD PDF]

... But could the different frequencies of "she" and "you" and "the" really be called "style"? On this, we disagreed. Some of us claimed that, though all styles do indeed entail linguistic choices, not all linguistic choices are however enough to speak of a style; others countered this argument by stating that, once an author or a genre opts for a certain linguistic choice, this is really all we need for our analysis, as a style follows necessarily from this fundamental level. This was the genuinely reductionist position – style as nothing but its components – and the more logically consistent one; the other position admitted that it couldn't specify the exact difference, or the precise moment when a "linguistic choice" turned into a "style", but it insisted nonetheless that reducing style to a strictly functional dimension missed the very point of the concept, which lay in its capacity to hint, however hazily, at something that went beyond functionality. Our job should consist in removing the haze, not in disregarding the hint.

4. A QUANTITATIVE LITERARY HISTORY OF 2,958 NINETEENTH-CENTURY BRITISH NOVELS: THE SEMANTIC COHORT METHOD
[DOWNLOAD PDF]

The nineteenth century in Britain saw tumultuous changes that reshaped the fabric of society and altered the course of modernization. It also saw the rise of the novel to the height of its cultural power as the most important literary form of the period. This paper reports on a long-term experiment in tracing such macroscopic changes in the novel during this crucial period. Specifically, we present findings on two interrelated transformations in novelistic language that reveal a systemic concretization in language and fundamental change in the social spaces of the novel. We show how these shifts have consequences for setting, characterization, and narration as well as implications for the responsiveness of the novel to the dramatic changes in British society.

This paper has a second strand as well. This project was simultaneously an experiment in developing quantitative and computational methods for tracing changes in literary language. We wanted to see how far quantifiable features such as word usage could be pushed toward the investigation of literary history. Could we leverage quantitative methods in ways that respect the nuance and complexity we value in the humanities? To this end, we present a second set of results, the techniques and methodological lessons gained in the course of designing and running this project.

FIGURE 1. Website of Literary Lab: litlab.stanford.edu

revelation, but nonetheless intriguing that differences are found at every structural level.

Although the content domain is quite different from our typical notion of Big Data, the research generated by the Literary Lab uses many of the now-familiar tools and analysis approaches: automatic feature extraction and data mining, clustering, principal components analysis, and networks. Matthew Jockers, a cofounder of the Literary Lab, has just published a book, *Macroanalysis: Digital Methods and Literary History,* that summarizes much of the current statistical work in this field and holds promise to bring awareness of the approach to a wider audience of literary critics. Since the area is still somewhat in its infancy, I think that there are opportunities for statisticians to bring our expertise to the table.

As an avid reader myself, it is easy to see both the potential and danger in Moretti's approach, as with much of Big Data. I say "potential" because, as Moretti and his colleagues are careful to point out, the statistical analysis is not meant as a replacement for reading, but rather a supplement. Bolstered by new analysis techniques, it is possible to uncover relationships among readers, writers, characters, and texts that were not previously imagined or discoverable. The technical reports are fascinating to read, and they raise some intriguing questions, such as how much of this is purposeful? For example, did Shakespeare know that Hamlet, Claudius, and Horatio, while all more or less equally entrenched in the play (in terms of density of connections to other characters), have different structural network roles? Was *Hamlet* designed so that almost all characters who interacted directly with both Hamlet and his uncle would wind up dead? How is it that gothic novels differ at the level of words—and not just the "obvious" thematic words (castles, wind-swept moors, . . .) but also filler words like "of " and "but"—the most commonly used words? Such questions expand our horizons as readers and appreciators of literature, urging us to think about the texts in new and exciting ways.

The danger creeps in at the point where someone—a teacher, a student, an administrator—decides that actually reading the plays or the books is no longer necessary. Simply dump them into a computer program, chug away, and voila!—instant analysis, if not comprehension or appreciation. Moretti and his team of researchers, I'd emphasize, do not at all consider their approach a replacement for reading but instead choose to focus on the opportunity to enhance enjoyment and understanding. Literature is probably by far the most common artistic domain to be subjected to date to statistical techniques, but there are other examples as well. At the website of the Museum of Modern Art, one can find the Connections tool as part of the Inventing Abstraction exhibit. The tool presents a connected network of artists, with Picasso and Kandinsky squarely in the center and highly connected. Individuals are nodes, and two nodes are connected if an acquaintance between the individuals in question in the years 1910–1925 (the focus of the exhibition) can be documented. The network tool is interactive; if you click on an individual artist, a new screen pops up with that artist's subnetwork of connections, some information about the artist, and thumbnails of representative works.

The first perspective—the use of statistical methods to analyze massive literary (and possibly other arts) data—is well established. The second perspective—the creation of art from data—ties in to a venerable tradition in statistics of data visualization but moves beyond principles of aesthetically pleasing presentations as advocated, for instance, by Edward Tufte and others. Rather, the modern artistic take on Big Data is on a grander scale. The Open Data Institute in London, for example, has installations that are data driven. One artwork, *The Obelisk*, scans online websites in real time and changes from opaque to transparent according to the number of references at a given moment to "crimes against peace" (as defined at the Nuremberg trials after World War II). Another, *Vending Machine*, releases free snacks when search terms related to the recent global recession make headlines on BBC News. The entire collection can be viewed at www.theodi.org/culture/collection.

Mark Hansen, a statistician who is a professor of journalism at Columbia University, has been working for many years at the interface between data and art, creating installations that depict communication—for instance, a piece in the lobby of *The New York Times* building that flashes phrases and words from the newspaper's vast database onto a grid of computer monitors. Phrases can be chosen according to some criterion, such as "ends in a question mark" or from searches and commentaries of the *Times* website, yielding the "pulse" of the paper. I was lucky enough to hear Mark talk about his work about 10 years ago and see firsthand examples of his innovative approach to representing massive amounts of data.

Finally, there is what I have called "arts analytics," for lack of a better name. In this perspective, artistic data, especially music, are treated like any other commodity and analyzed. Who will be the next big act? If a listener likes artist A, what other artists is he or she likely to enjoy? There are massive data sets available for this type of analysis, collected by Spotify and other online streaming music sources. The record company EMI took this approach even further and, in July 2012, created the Million Interview data challenge and "hackathon," a massive project to develop methods for predicting song popularity. For the purposes of the challenge, EMI collected information about interests, attitudes, familiarity, and appreciation of music fans worldwide. Teams and individuals from around the world were then able to use the data set to devise and test algorithms for the prediction of musical taste. The winning team, from a data mining company in China, used common statistical techniques, such as the singular value decomposition and logistic regression. Other teams used decision trees and random forests. The website nextbigsound.com uses information from social media (Facebook "likes," etc.) to predict popular music acts. This is a website geared toward the artists themselves, with the stated goal of helping musicians make decisions about their careers and record labels make decisions about marketing, production, and so on.

The concerns about Big Data in the arts come heavily to the forefront when we talk about arts analytics. It's one thing to worry about computers replacing humans as readers—will this ever really happen? But it's quite another, in my opinion, to attempt to quantify and predict tastes and preferences in music or the other arts. We've already, to a great extent, lost that wonderful experience of discovering a new writer by browsing the shelves of a bookstore, or a new singer by listening to the radio. New technologies and new ways of marketing art to the general public mean that we can (and many do) isolate ourselves in our predetermined bubbles. Do I want a computer algorithm telling me what music to listen to? It's a double-edged sword! If that algorithm introduces me to performers and genres I might not have found on my own, then certainly, bring it on! But if it serves merely to reinforce my current tastes, I'm not so sure this is something I would sign on for. On the other hand, in the course of researching this column, I came across a band called "Big Data"—check them out at bigdata.fm. As described on their website, they are an electronic duo that aims to explore the relationship between man and machine through their music, as well as how the Internet has changed the human experience. The music expresses a deep cynicism and distrust of our growing dependence on technology and computing—a reassuring expression of the artistic impulse.

As you have probably noticed if you have been following the conversations about Big Data in both the popular press and our professional publications, along with the excitement of this new field and approach, there is some fear; indeed, some have already written about a "Big Data backlash." Part of this phenomenon flows from misgivings about the intrusiveness of data collection and observation in our daily lives; part stems from a feeling that this "next big thing" has already (!) failed to live up to its potential—that Big Data and its tools were somehow going to solve all the problems of this or that industry, an unrealistic expectation to say the least. In all of this, there also has somehow been the feeling that the arts are—or should be—different, isolated or aloof from the supposed coarseness of a data-driven approach. The backlash against data analysis more generally, and Big Data specifically, in this context then sits against the background of wanting to preserve some element of our existence that is off limits. I can understand and sympathize with this sentiment to a large extent. But I can also see the value of the many and growing intersections between art and statistics. I can't read every book written, and even if I could, I probably wouldn't be able to discover certain connections and themes that unite *and* differentiate between genres and authors. An algorithm can do this easily—and thereby open up new questions for me to explore as a reader. It's interesting to learn that painters who try to copy the style of van Gogh need to use more brushstrokes to achieve certain effects—not a surprising finding, one may rightly claim, but still one that opens up possibilities for further exploration. And a band called "Big Data," while worried about the ramifications of data and technology in our lives, can still get my family dancing around the kitchen on a Sunday afternoon!

Further Reading

winedarksea.org. A website about text analysis from a Big Data perspective.
litlab.stanford.edu. Website of the Literary Lab; pamphlets available for download.
nyuisva.wordpress.com. A website about information systems in the visual arts with the slogan "using the web to serve culture."
Jockers, M. L. 2013. *Macroanalysis: Digital Methods and Literary History.* University of Illinois Press.

On the Number of Klein Bottle Types

Carlo H. Séquin

Introduction

A Klein bottle is a closed, single-sided mathematical surface of genus 2, sometimes described as a closed bottle for which there is no distinction between inside and outside. A canonical example of this surface is depicted in Figure 1A, with its characteristic "mouth" at the bottom where the "inside" of the surface connects to its "outside." However, if one consults Wikipedia or some appropriate math text, one can find Klein bottles that look radically different, such as the twisted loop with a figure-8 profile (Figure 1B) or Lawson's [9] single-sided surface of genus 2 (Figure 1C), which has been conjectured to minimize total bending energy for such a surface [8]. Probing further, one can also find the intriguing glass sculptures of Alan Bennett [3] (Figure 2) and various forms of knotted Klein bottles presented at Bridges 2012 [15], some of which are depicted at the end of this article (Figures 14 and 15).

Clearly, all these shapes are geometrically different. But how many of them are *fundamentally* different, if smooth deformations are allowed to transform one shape into another? If we ask mathematicians how many different types of Klein bottle there are, they typically say, "Three or four—it all depends on whether the surface of the bottle is *marked*, e.g., decorated with a coordinate grid." To come to this conclusion, mathematicians allow smooth deformations that are called *regular homotopies*. These are transformations in which a surface is allowed to pass through itself; but the surface may not be punctured, cut, or torn, and it must not assume any sharp creases or form any other singular points with infinitely sharp curvature. The precise conditions that a regular homotopy must satisfy are even more stringent [6, 7, 10], but those subtleties have no effect on the discussions that follow. For well-versed mathematicians, this paper may not break novel ground. All of the topological facts and insights are contained in previous work by others [1, 2, 6, 10]. The goal of this paper is to make these important findings accessible and understandable to a much broader audience and to introduce representatives of the different Klein bottle classes in as symmetrical and easy-to-understand a form as possible (Figure 13). To set the stage, we start with a discussion of the regular homotopy classes of annuli and of Möbius bands, which then enable the construction of the three types of unmarked Klein bottles; two are chiral and mirror images of each other, and one is amphichiral. We then show that when surface markings are introduced, the latter class splits into two separate regular homotopy classes. We conclude with a presentation of several more complex Klein bottles and outline an analysis technique that should allow the reader to classify those contorted geometrical shapes as well as Bennett's sculptures (Figure 2).

Regular Homotopies of Twisted Bands

We first analyze how many different types of twisted bands can be formed that cannot be transformed into one another via regular homotopies. We start with the rectangular piece of surface depicted in Figure 3A. Both sides are decorated with two chains of white and black arrowheads to provide a sense of directionality. The "front" face also shows a rainbow-colored background, and the "back" side exhibits a checkerboard pattern. Now we explore different ways of closing this band into a loop.

First we join the two short, vertical edges without imparting any twisting motion; this will form an untwisted annulus or generalized cylinder with two clearly distinct surfaces: one is rainbow-colored and the other is checked (Figure 3B). Alternatively, we could apply a half-twist of 180° before joining the two ends (Figure 3C). This manipulation leads to a surface that is clearly quite different. At the junction, the rainbow pattern joins the checkerboard; thus we can travel from one face to the other without having to climb over the rim or boundary of the surface. The surface is now said to be single-sided: it is a Möbius band. This surface is topologically different from the annulus, and these two shapes cannot possibly be transformed into one another; to do so would require changing the appearance of the marked line in

FIGURE 1. Klein bottles with different geometries: A) Cliff Stoll with a very large classical Klein bottle (from the artist's web page [18], used by permission of Cliff Stoll); B) woven figure-8 Klein bottle (by C. H. Séquin 1997); C) Lawson's minimum-energy surface approximated with a fused-deposition model (by C. H. Séquin 2011).

Figure 3B to that in Figure 3C, i.e., changing the structure of the underlying surface.

But there are more options. We could apply one or more full twists of 360° before joining the two ends. These surfaces are then all two-sided. But are they all in the same regular *homotopy class*? Conversely, if we apply an odd number of 180° half-twists, the surfaces are always single-sided, but not all of them can be transformed into one another. Furthermore, we could let the band form a figure-8 shape (Figure 4) or pass through multiple loops before we join the ends (Figure 5). Before we can classify all these many possible shapes, we have to understand how the apparent twist in the band is affected as we try to unwind a more complicated path through space into a simple loop with a planar, circular *core curve* (the center line of the band).

UNWARPING A FIGURE-8 PATH

Since we are not allowed to form any sharp creases or pinch-off points, we can only transform a figure-8 path (Figure 4A) into a circular path by making a move through three-dimensional (3-D) space (Figure 4B–E). We start with an initially "untwisted" band, where the amount

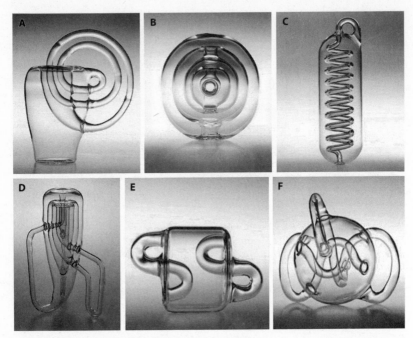

FIGURE 2. Glass sculptures by Alan Bennett: (A, B, C) fancy Klein bottles exhibiting artistic geometry. More complex shapes: D) three nested Klein bottles; E) two-handle Klein bottle; F) three-handle Klein bottle. All models are at the Science Museum/Science & Society Picture Library in South Kensington, U.K. [3], used by permission.

of *twist* is determined by the amount of extra rotation that needs to be applied to a rotation-minimizing frame in order to keep the band closed smoothly. As the initially planar, figure-8-shaped core curve is deformed into a warped space curve, a steadily increasing amount of twist has to be added to keep the band smoothly closed. By the time the core curve has been opened up into the final, circular loop with again a planar core curve, a twist of 360° has been imparted onto the ribbon (Figure 4E). Conversely, an undesirable twist of ± 360° can be converted into a figure-8 sweep path. Furthermore, if we start from Figure 4E, first warp the band into a figure-8 shape (Figure 4A), then continue the process by letting the band pass through itself until the core curve exhibits again a perfectly circular loop, we find that the total twist in the band changes by exactly 720°. Thus the twist in any closed ribbon can readily be changed by $\pm\,720°$ by performing such *a figure-8 sweep cross-over move*. This fact is closely related to Dirac's *Belt Trick* or Feynman's *Plate Trick* [5]. It also played an important role in a similar study [13, 16] to find all the different regular homotopy classes of topological tori—two-sided, orientable surfaces of genus 1. It turns out to be true in general that topologically, the twist in any such band should be counted modulo 720°.

UNWINDING AN EXTRA LOOP

If the initial configuration exhibits multiple loops (Figure 5A), we can unwind such a coil in the following way: one loop at a time is folded out and swung like an opening door through 180°, as is shown for the outermost loop in Figure 5B–D. The resulting figure-8 configuration may still have multiple loops in one of its lobes (Figure 5D). Note that this folding-out process by itself does not change the twist in that loop, even though it does change the azimuth and switch the "inside" and "outside" of the moving ribbon segment. (The amounts of twist imposed on the upper and lower half of the moving loop because of the 3-D space-warp of the corresponding core curve segments cancel each other out exactly.) However, the unwarping of the figure-8 lobe generated by the unfolding process (using the process depicted in Figure 4) adds a twist of $\pm\,360°$, as discussed above.

Of course, the starting configuration may be a more complex tangle, in which the ribbon may be twisting through three dimensions, may be knotted, or may even be selfintersecting. However, in the domain of regular homotopies, we are able to move portions of this ribbon smoothly through each other. We can thus readily untangle any knot and always bring the core curve into a planar, circular configuration. In this state, we can now unambiguously determine the twist of the ribbon by recording how the surface normal of the ribbon rotates with respect to the Frenet frame of the core curve. If the integral amount of rotation exceeds the range $[-180°, +360°]$, then we can reduce the amount of twist in increments of $\pm\,720°$ through the application of the figure-8 sweep cross-over move depicted in Figure 4.

We thus find that for all closed, twisted bands, there are two topological equivalence classes (single-sided and double-sided) and two regular homotopy classes in each of them. They are best represented

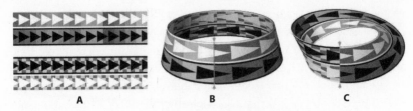

FIGURE 3. A) Rectangular domain, front and back. B) Merging the two short edges to form an untwisted annulus. C) Closing the loop after a left-handed half-twist of 180° has been applied to the band. See also color images.

FIGURE 4. Unwarping a figure-8 path: A) starting shape; B) warped space curve, 90° twist; C) seriously warped curve, 180° twist; D) 320° twist; E) circular loop with 360° twist. See also color images.

by the untwisted annulus, designated **AO** (Figure 3B), by an annulus with ± 360° of twist, labelled **AT** (Figure 4D), and by a left-twisting Möbius band, **ML** (Figure 3C) and its mirror image—a right-twisting Möbius band **(MR)**. Note in particular that none of the regular homotopies that we have examined so far can transform a Möbius band, say **ML,** into its mirror image **(MR)**; in fact, it can be proven formally that no regular homotopy can perform such a transformation [6, 7]. These important facts will now be used in the analysis of the number of Klein bottle types.

Constructing Klein Bottles

A Klein bottle is a single-sided surface of Euler characteristic $\chi = V - E + F = 0$ with no perforations or boundaries. It is essentially a single, contorted, self-intersecting "tube." It has been proven that in

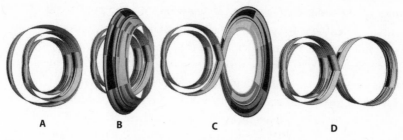

FIGURE 5. Unwinding an extra loop: A) starting configuration with planar core curve; B) 45° worth of unfolding; C) 90° worth of unfolding; D) final figure-8 loop, again with a planar core curve. See also color images.

Euclidean space $\mathbf{R}^3$, no *embeddings* of Klein bottles are possible—only *immersions* that exhibit some self-intersections [11]. Several different kinds of possible immersions will now be described through simple geometrical constructions of some basic Klein bottle types.

The Classical "Inverted-Sock" Klein Bottle

We start with the rectangular domain depicted in Figure 6A. Since we want to construct only single-sided surfaces, we assume that this surface is made of a thin "glassy" material that carries the desired texture within, so that the two views from above and below a small patch of the surface will appear to be mirror images of the same pattern. The crucial point to forming a Klein bottle is to join opposite sides of this rectangle while observing the directionality indicated by the four arrows along the rectangle sides. For an illustrative construction, we first join the two longitudinal (horizontal) edges (marked with arrows above and below in Figure 6A to form a generalized tube. This tube may have a simple circular cross section; alternatively, it may assume a self-intersecting figure-8 profile or even form a more complicated, multi-rolled tube. The difficult step is to now join the remaining open edge pair (marked with arrows next to the numbered sides of the rectangle in Figure 6A with reversed directionality. Different types of tubes require different deformations to accomplish this.

For the classical, "inverted-sock" Klein bottle with an **O**-profile, the closing of the loop is most conveniently done by narrowing one end of the tube and inserting it sideways into the larger end of the tube—forming a **J**-shaped loop (Figure 6C); this configuration properly lines up all the numbered labels. Now the two concentric ends are merged by turning the smaller one inside out. This step yields the classical "inverted-sock" Klein bottle, named **KOJ** (Figure 7A). In this case, the lower half and the upper half of this merged loop independently form two Möbius bands of opposite handedness, as shown in Figure 7B and 7C. We can denote this relationship symbolically as **KOJ = MR + ML**.

FIGURE-8 KLEIN BOTTLES

When merging the two horizontal edges marked by the top and bottom arrows in Figure 6A to form an initial tube, we are not forced to form a round, circular **O**-profile. Instead, we may form a figure-8 cross-section (Figure 8A) or an even more complicated, multiply rolled generalized cylinder, as was discussed for tori [14]. These alternatives result in various Klein bottles, some of which may belong to different regular homotopy classes and thus could not be smoothly deformed into the classical "inverted-sock" shape.

For the tube with the figure-8 profile, there are a few different ways in which we can fuse the tube ends with the required reversed orientation, and they may result in different Klein bottles. For instance, we can bend the tube into a simple toroidal loop (an **O**-shaped path) and give the figure-8 cross section a 180° torsional flip (Figure 8B); this flip can either be clockwise (right-handed: **R**) or counterclockwise (left-handed: **L**). These two options result in two figure-8 Klein bottles that belong to two different regular homotopy classes, which we call **K8R-O** (Figure 9) and **K8L-O** (Figure 10), respectively. To see that there can be no regular homotopy between these two types, assume that we remove the light yellow area from the rectangular domain shown in Figure 6A, and focus our attention on the remaining light blue area. This domain happens to be a right-handed Möbius band **MR** in the Klein bottle of type **K8R-O** (Figure 9), but a left-handed Möbius band in the Klein bottle **K8L-O** (Figure 10). As mentioned above, no Möbius band can be transformed smoothly into its mirror image. The darker region and the lighter region of the texture domain (Figure 6A) appear in each of **K8R-O** and **K8L-O;** Figures 9 and 10 show these

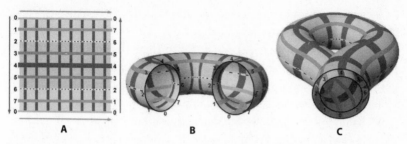

FIGURE 6. Construction of a Klein bottle: A) its rectangular domain; B) the domain rolled up into a tube; C) the tube bent into a J-formation with its two ends lined up so that labels match. See also color images.

regions as individual Möbius bands, as well as joined together into Klein bottles. With this particular partitioning of the original domain, the pairs of Möbius bands that form the Klein bottles happen to intersect along their center lines. The domain could also be split in a different way, e.g., along a diagonal line; such a split would still partition the surface into two Möbius bands of equal chirality.

YET ANOTHER FORM OF KLEIN BOTTLE

Figure 11A shows a third way by which the figure-8 profile tube can be closed into a Klein bottle. This new type of Klein bottle is constructed by forming the same kind of J-shaped sweep path as for the classical Klein bottle and fusing the two nested figure-8 profiles by turning one of them inside out. To do this smoothly, we use an asymmetrical figure-8 profile in which one lobe is larger than the other one. As this profile is swept from one end of the tube to the other one, the larger lobe shrinks and the smaller one grows, so that the end profiles can be nested as shown in Figures 8C and 11A. Now a nicely rounded figure-8 end cap can smoothly close off this Klein bottle (Figure 11B). I have not seen this particular Klein bottle depicted previously. It is rather special, since it does not just have a single self-intersection line like all the other models, but features two triple points. These triple points occur where the self-intersection line of the figure-8 cross section passes through the wall of the other tube near the mouth of the Klein bottle. Banchoff has shown that when triple points occur in Klein bottles, they must always occur in pairs [1].

A closer analysis of this Klein bottle reveals that its two constituent Möbius bands are of the same handedness, and it therefore comes in two different chiral varieties; we refer to them as **K8R-JJ** and **K8L-JJ**, respectively. We can split this Klein bottle cleanly along the closed self-intersection loop passing through both triple points to obtain two Möbius bands. Each such component on its own resembles an "inverted-sock" Klein bottle shape with a sharp crease line, which itself exhibits a Möbius twist (Figure 11C). As we will see shortly, these type **K8x-JJ** Klein bottles (i.e., **K8L-JJ** and **K8R-JJ**) are also in the same two regular homotopy classes as the **K8x-O** bottles, respectively.

Marked Klein Bottles

A paper by Hass and Hughes [6] states (p. 103):

> Corollary 1.3 (James-Thomas): There are 2^{2-x} distinct regular homotopy classes of immersions of marked, closed surfaces of Euler characteristic χ into $\mathbf{R}^3$.

This corollary tells us that there should be *four* distinct, marked Klein bottle types that cannot be transformed smoothly into one another when surface markings are considered. Although Hass and Hughes give complete recipes for how to construct representatives of each type, I have found no paper that shows explicit pictures of all four types. The single-sidedness of these objects also makes it conceptually more difficult to visualize these shapes.

Since regular homotopy transformations cannot turn a left-twisting Möbius band into a right-twisting one [6, 7], we can immediately identify three types of Klein bottles that differ based on the handedness of the two Möbius bands from which they are composed, and which thus must belong to three different regular homotopy classes. The simplest representatives of the three classes are: **K8L-O = ML + ML; K8R-O = MR + MR;** and **KOJ = ML + MR.** The first two are chiral, and **KOJ** is amphichiral (its own mirror image). These three classes are *structurally* different; they do not depend on any markings on the surface. This fact reveals an interesting difference from the world of tori [13, 14], where there are only *two* structural classes: one formed by the three tori types **TOO, TO8** and **T8O,** and the other one by **T88** by itself.

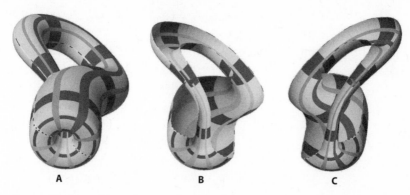

FIGURE 7. The ordinary "inverted sock" Klein bottle resulting from the construction in Figure 6: A) the complete Klein bottle **KOJ**; B) its lower half is a right-handed Möbius band (**MR**); C) the upper, left-handed Möbius band (**ML**) is shown rotated 90° to the right. See also color images.

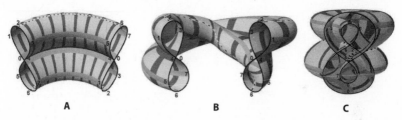

FIGURE 8. Constructions for figure-8 Klein bottles: A) figure-8 tube; B) the tube twisted through 180° so that the two ends can be merged into a toroidal loop; C) a new way to line up the number labels. See also color images.

FIGURE 9. Two right-handed Möbius bands **MR** (A, B) form a right-handed Klein bottle of type **K8R-O** (C). See also color images.

Now, where do we find the fourth anticipated type of Klein bottle? It cannot be composed of two Möbius bands of the same handedness because if we could find an additional bottle of this type, then its mirror image would be equally novel, and we would end up with five different types of Klein bottles. This issue also tells us that Lawson's Klein bottle, **K8x-Lawson** (Figure 1C), as well as the "inverted double sock," **K8x-JJ** (Figure 11B), both of which display C_2-symmetry and an explicit handedness, cannot provide the sought-after fourth representative and that there must be a way to transform them smoothly into the corresponding simpler twisted figure-8 loops, **K8x-O** (Figures 9 and 10). Thus, the sought-after fourth type has to be composed of two Möbius bands of *opposite* handedness.

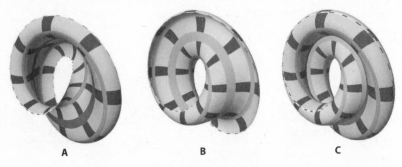

FIGURE 10. Two left-handed Möbius bands **ML** (A, B) form a left-handed figure-8 Klein bottle **K8L-O** (C). See also color images.

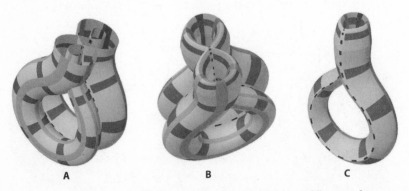

FIGURE 11. A Klein-bottle based on an "inverted double sock": A) without end caps to show the nesting of the tube profiles; B) completed Klein bottle of type **K8L-JJ**; C) one of its two **ML** Möbius bands. See also color images.

We can try to start from **KOJ** and somehow "switch" the two Möbius bands. But if we try this, and in this process flip the handle in Figure 7A to the right-hand side, we can restore the original state by rotating the whole bottle through 180° around the axis coming out of its mouth—unless we pay attention to the coloring! If the original **KOJ** is composed of a lighter colored **MR** and a darker colored **ML**, then its mirror image comprises a light-colored **ML** and a darker colored **MR**, and no regular homotopy can deform one into the other.

Adding Collars to a Klein Bottle

Topologists also have some other techniques for generating new representatives of a surface that may then fall into different regular homotopy classes. Hass and Hughes [6] start with a closed curve on the surface, turn it into a crease, and then add a fold that forms a *kink* or *collar* along that line. Let us see what results we get from grafting such collars to a Klein bottle. On Klein bottles, because of their single-sidedness, adding collars takes some extra care. To understand this fact, we need to first discuss the various types of lines that can be drawn on a Klein bottle.

First, there are the *meridians*, shown as transversal bands in the examples in this article. If we remove such a meridional band from a Klein bottle, the surface left over has the topology of an open cylinder with either an untwisted **O**-profile or a (twisted) figure-8 cross section; these shapes are in the same regular homotopy classes as the annuli **AO** and **A8**, respectively. Second, there are the center lines of the two Möbius bands, which are single-sided and either left- or right-twisting. Next, there are all the other longitudinal lines that run on either side of the Möbius core curves; they run twice around the loop before they close onto themselves, and they have an integer amount of twist. The removal of a small strip of surface along such a line partitions the Klein bottle into two Möbius bands. Finally there are diagonal lines that can be seen as connected combinations of the aforementioned lines. And, of course, there are trivial local loops that can be contracted to a single point; they are of no interest in our discussions and should be avoided.

We can graft a meridional collar anywhere along the tube forming the Klein bottle. In particular, we can use the meridian at the rim of the mouth (Figure 12B), which then results in the collared structure **KOJ-C** shown in Figure 12C. The fold of this graft forces the tube to

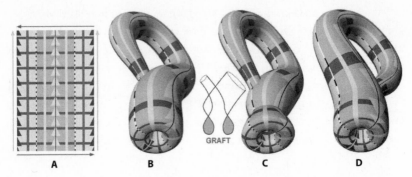

FIGURE 12. A) New texture with longitudinal directionality; B) classical **KOJ** with this texture applied; C) **KOJ** with a collar grafted at the mouth; D) **KOJ** with inflated left branch, eliminating the collar self-intersection, and resulting in a swap of the two colors and reversed longitudinal directionality. See also color images.

undergo an extra surface reversal and thus changes the directionality by which the directional texture (Figure 12A) passes through the mouth. Whereas in Figure 12B the arrows seem to flow out of the mouth, in Figure 12C they flow *into it*. It turns out that we can easily eliminate the newly generated circular intersection line created by the grafted-on collar by inflating the left branch of the tube; we then simply obtain a Klein bottle in which the thin and the thick arms emerging from the mouth have been switched. But both Möbius bands now exhibit reversed handedness! Structurally, this makes no difference, since the original **KOJ** could simply be rotated 180° around the mouth axis to coincide with the new shape. However, if we place a texture on the surface that exhibits directionality in the longitudinal direction or which colors the two Möbius bands differently, then the result is indeed in a different regular homotopy class. Here is an analysis of the twistedness of the various characteristic ribbons in response to the added meridional graft: the meridional ribbons are not affected. But the longitudinal Möbius bands, which pass through the grafted collar exactly once, experience an extra twist of $\pm$ 360° that reverses their chirality; the left-twisting Möbius band becomes right-twisting and vice versa. This phenomenon confirms that we have found a representative of the fourth regular homotopy class for Klein bottles (Figure 12C and 12D); this is my preferred representation of this type.

We can also graft a fold onto a meridian of a Klein bottle with a figure-8 cross section. Such a collar travels on the outside for half the figure-8 and on the inside for the other half of the profile and then closes smoothly onto itself. Again, the meridional bands are not affected. Each Möbius band crosses this fold exactly once and thus changes its twist by 360°, which reverses its twistedness. Thus, adding a figure-8 shaped M-graft will turn **K8R-O** into **K8L-O** and vice versa, but it does not generate a new regular homotopy class.

Adding a collar along a general longitudinal line that passes around the loop twice does not affect other longitudinal lines. It adds two reversing lobes to every meridian and thus does not affect their net twist. Such L-grafts thus do not change the regular homotopy class. A more complete discussion of all the collar-grafting options and how they convert Klein bottles from one regular homotopy class into another one can be found in a separate report [17].

Change of Surface Parameterization

For a torus, some discrete changes in the parameterization grid may yield a torus that belongs to a different regular homotopy class. In particular, the introduction of a Dehn twist [4] of 360° in the meridional direction (M-twist) converts a torus of type **TOO** into **TO8**, and a longitudinal (or "equatorial") Dehn twist (L-twist) takes **TOO** into **T8O**. Thus, we can try to apply all the surface reparameterizations that we have studied for the case of tori [13, 14] and see how they affect the classifications of various Klein bottles. However, we already know that none of them can change the structural classification since the effect of such reparameterizations would be completely invisible on unmarked glass surfaces.

Again the situation is somewhat tricky on Klein bottles. Because of their single-sidedness, meridians cannot be assigned a consistent direction. This property rules out the possibility of a parameter swap that switches the roles between meridians and longitudes. On the other hand, it makes surface "eversion" trivial, since we can just shift the surface texture by one period in the longitudinal direction to have any surface normal point in the opposite direction. Also, as we have seen above, simply reversing the directionality in the longitudinal direction puts a **KOJ** Klein bottle into a different regular homotopy class; but such an L-reverse operation has no effect on the **K8x** types.

Introducing Dehn twist on Klein bottles also demands some extra care. Adding 360° of meridional Dehn twist (M-twist) to a **KOJ** bottle is equivalent to adding a meridional collar: one feature can be turned into the other one by applying Dirac's belt trick [5] as illustrated in Figures 23 and 24 of my report on tori [16]. Alternatively, we can remove any amount of twist in the handle of a **KOJ** type by swinging the handle through half the twist-angle around the mouth axis of the bottle [12]. On a marked **KOJ** bottle, such a handle rotation shifts longitudinal curves out of alignment for all angles except for rotations in multiples of 180°. Keeping the coloring fixed on the outer, thicker arm near the mouth and swinging the handle around through 180° transforms the Klein bottle into its mirror image and thereby reverses the chirality of both of its Möbius bands. The fact that each Möbius band changes its chirality as we apply an M-twist of 360° can readily be understood when we contemplate what happens to the one end of a Möbius band, cut at a circular meridian, that is then forced to slide around this circle (which has a turning number of 1).

With the same mental imagery, we can also see that applying an M-twist to any bottle of type **K8x** does not change the twistedness of the two Möbius bands since in this case the figure-8 shaped meridians have a turning number of zero.

For completeness, we should also investigate the effects of Dehn twists along longitudinal and diagonal loops. It turns out that such operations must be disallowed on Klein bottles since they lead to inconsistencies in the parameterization [17].

Map of the Regular Homotopy Classes of Klein Bottles

After having examined several different schemes for constructing Klein bottles and having identified the anticipated four different regular homotopy classes, we can now draw the complete map of all Klein bottle types, as was done previously for tori [13, 14]. Any possible Klein bottle, no matter how weird or how twisted, can be deformed by a regular homotopy into one of the four simple representatives depicted in Figure 13. In summary, we see that there are now three structural domains, one of which splits into two parts when surface parameterization is taken into account.

For each case, a small picture shows my preferred representative for that class, and next to it, a symbolic representation of four characteristic ribbons on that surface. The transversal vertical stripe represents the meridional bands, and the horizontal stripes denote the center lines of the two Möbius bands as well as a general longitude ribbon that separates the two bands. The labels next to them tell whether that band is untwisted (**O**) or fully twisted (**8**), and whether a Möbius band is left-twisted (**L**) or right-twisted (**R**).

Outside the grey circles there are some lines with double arrows, indicating various transformations that could be applied to a coordinate grid on the surface of the Klein bottle and how they would affect the regular homotopy type of the transformed surface. As discussed above, the addition of a meridional collar or equivalently, the addition of 360° of meridional twist keeps the structure of the **KOJ** bottles the same but changes the regular homotopy class for a marked surface; it either reverses the directionality in the longitudinal direction or it switches color between the two Möbius bands. Adding a meridional collar to a **K8L** bottle turns it into a **K8R** bottle and vice versa. For a tube with a figure-8 profile, this is not equivalent to adding a meridional Dehn twist; the latter leaves the **K8x** types unchanged. Also, a reversal of the longitudinal direction can readily be undone, by looking at a **K8x-O** toroid from the other side.

We also may wonder what the operation of *surface eversion*, which—somewhat surprisingly—kept all tori types within their own regular homotopy classes, would do for Klein bottles. It works the same way for Klein bottles, but here it is much easier to accomplish: we simply need to shift the coordinate grid one full period in the longitudinal direction, and every point will have moved to the "opposite" side of the surface.

On the other hand, the operation of *parameter swap*, which exchanges the roles of the meridional and longitudinal coordinate lines, is not possible for Klein bottles. All meridional lines always have an integral amount of full twists, whereas two of the longitudinal core curves exhibit the Möbius half-twists. No smooth continuous deformation can accomplish such a fundamental change.

SURFACE ANALYSIS

Figure 13 also tells how we can determine the types of the novel, fancy, twisted Klein bottles presented in the last section of this article. Those gridded models, as well as the glass sculptures by Alan

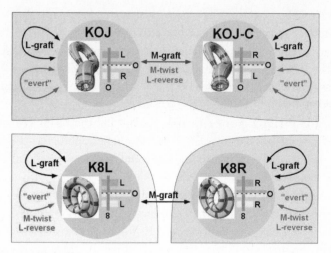

FIGURE 13. Complete map of the structurally different Klein bottle classes and of the effects of any surface reparameterizations. See also color images.

Bennett [3] depicted in Figure 2, have no indication of directionality on their surfaces; thus, we only need to figure out into which structural class they belong. First we make sure that the surface is indeed single-sided. For this we just need to find some arbitrary path that leads from a particular point on this surface to the "opposite side" at the same location. Next we check that the surface has Euler characteristic $\chi = V - E + F = 0$, i.e., it is essentially a single, contorted, self-intersecting "tube," possibly with some "inverted-sock" turn-backs—but with no branching. If this "tube" cannot be readily identified, we would have to place a coarse graph of edge-connected vertices onto the surface and explicitly count the number of vertices (V), edges (E), and facets (F) in that mesh.

Once we have established that the surface indeed has the topology of a Klein bottle, the next crucial step is to find a Möbius ribbon on that surface. This may be a little tricky on surfaces such as the ones shown in Figure 1C or Figure 15B. In those cases, we may want to not just trace out a single core curve but also accompany it by "parallel" side lanes on either side. As we trace this "highway" over the surface until it joins itself again, we may find that all three lanes join themselves; this means that we have traced out an orientable ribbon. We then need to continue our search until we find a "highway" loop in which the two side lanes merge into each other, indicating a true Möbius ribbon. If we then "subtract" that ribbon from the surface, what is left over forms another Möbius band. We now determine the twistedness of both of these two Möbius bands, and these observations define the regular homotopy structure of this Klein bottle as already pointed out in the section on Marked Klein Bottles: **ML + ML → K8L; MR + MR → K8R, and ML + MR → KOJ.**

To determine the twistedness of a space curve embedded in a complicated surface, I have found success with the following method: try to model in 3-D space a thin ribbon made from stiff paper so that it conforms to the Möbius band found on the surface in question (which may exist in the form of a 3-D model or simply as a 2-D rendering). When the paper strip has been shaped along the whole pathway of the ribbon, tape its two joining ends together, and then try to open it up into a circular loop in which the amount of twisting can readily be determined. In more complex cases, such as those depicted in Figure 14, the paper strip may actually be knotted. In this case, you have to cut and rejoin segments of this strip to untangle the knot; make sure that you preserve the alignment of the ends of the paper strip near those temporary cuts, so that the built-in twist is not altered. The total amount of twist observed in this circular loop can then be reduced modulo 720° to fall in the interval [−180°, +360°] to make classification easy.

It is quite possible that when we search for that first Möbius core curve, we might not stumble upon the most direct closure for such a curve and might find instead a line that makes an additional meridional loop around the "handle" of the Klein bottle, thus following some "diagonal" line. This loop may impart an extra twist of 360° on such a ribbon and thus may reverse the chirality of that Möbius band. Would this not lead to a different classification? No, it does not! This particular choice of a first Möbius band would also force the same extra meridional loop onto the second Möbius band. Thus in the case of a **KOJ** bottle, both Möbius bands would experience that extra twist and the resulting reversal of their chirality; thus the structural classification would remain unchanged. In the case of the **K8x** Klein bottles, the meridional band has a turning number of zero, and extra meridional detours do not add any twist to the Möbius bands; again, the classification remains unchanged.

Visualization Models and Sculptures

Mathematical visualization models are important to render math insights understandable to an audience that extends beyond mathematicians trained in topology and group theory. Making such visualization models as clear and unambiguous as possible is itself an art. Making them also aesthetically pleasing may entice a larger audience to pay attention and to study them. In some cases, well-conceived math models may even serve as inspiration for constructivist sculptures. In this section, I show some maquettes of gridded models of Klein bottles that might be turned into large sculptures by assembling them from I-beams or from tubular elements.

Figure 14 depicts two knotted versions of Klein bottles. The first one is a twisted figure-8 loop; however, in this case the loop undergoes three flips of 180°, and the sweep path is closed into a trefoil knot. Figure 14B depicts a different kind of trefoil knot; this one is composed of three interlinked segments of the classical "inverted-sock" type of Klein bottle. A surface reversal takes place at each one of the three Klein bottle mouths. As long as there is an odd number of such surface reversals, the result is a Klein bottle, rather than a warped self-intersecting torus. I like to call such knotted variants of Klein bottle shapes *Klein knottles* [15].

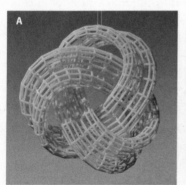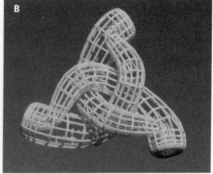

FIGURE 14. Maquettes of knotted Klein bottle sculptures: A) Trefoil knot with figure-8 profile; B) trefoil formed by three Klein bottle sections of type **KOJ**.

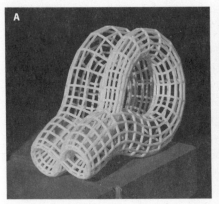
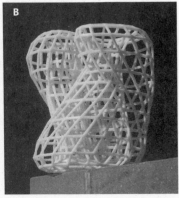

FIGURE 15. Unusual, heavily self-intersecting Klein bottle structures: A) The left-turning **K8L-JJ**; B) a right-turning Klein bottle with figure-8 profile and a zig-zag sweep path, **K8R-ZZ**.

Figure 15 displays some novel, more radical shapes for Klein bottles. First there is a gridded version of the "inverted double sock" Klein bottle, **K8L-JJ**, already introduced in Figure 11B. This gridded, "semitransparent" model allows a better inspection of how the figure-8 cross section turns inside out at the Klein bottle mouth and then morphs from one asymmetrical configuration to another one, in which the uneven sizes of the two lobes are reversed. As in the classical Klein bottle, the figure-8 profile mouth also accomplishes a surface reversal. Thus, any untwisted configuration with an odd number of such reversals topologically forms a Klein bottle. However, the figure-8 profile gives us another degree of freedom. It allows us also to accomplish a surface reversal by giving the figure-8 tube a 180° flip—or an odd multiple thereof. These two mechanisms of surface reversals can be combined: we may use an even number of 180° flips together with an odd number of Klein bottle mouths, or alternatively, an even number of turn-back mouths together with an odd number of flips. Figure 15B displays a very "compact" Klein bottle geometry, **K8R-ZZ**, with two Klein bottle mouths, combined with an overall twist of 180°. This twist is applied in two portions of 90° each, as the figure-8 profile travels once up and down a portion of the z-axis. In one of these two passes, the profile also morphs into its twin shape, in which the sizes of the two lobes are exchanged.

Conclusion

For most people who learn of the existence of Klein bottles, even if it is just the most frequently depicted "inverted-sock" geometry, these are fascinating objects. When one learns that there are several different Klein bottles that are not just smooth deformations of one another, these surfaces become even more interesting. Next, one may come across somewhat contradictory statements that claim that there are three or four "different" Klein bottle types. Typically, these claims are not accompanied by any explicit pictures of what these types may look like, and now one faces a really intriguing puzzle. Unfortunately, the mathematics literature that contains precise answers to all these issues is of a very technical nature, and thus the answers to these interesting questions have remained inaccessible to many potentially interested readers. I hope that this article sheds some light on these issues and further expands the catalogue of intriguing geometries that constitute topological Klein bottles.

Acknowledgments

I would like to express my thanks to Dan Asimov, Rob Kusner, and Matthias Goerner for insightful and constructive comments on preliminary versions of this paper. I am truly indebted to the two anonymous reviewers whose extensive comments helped me to reduce ambiguous or misleading wordings and to enhance the clarity of the presentation. Their suggestions also encouraged me to provide better and more detailed explanations, and in this process I gained a deeper understanding of some of the underlying mathematical concepts.

This work was supported in part by the National Science Foundation (NSF award No. CMMI-1029662 (EDI)).

References

[1] T. F. Banchoff, "Triple points and surgery of immersed surfaces." *Proc. Am. Math. Soc.* 46 (1974), 407–13.

[2] T. F. Banchoff and N. H. Kuiper, "Geometrical class and degree for surfaces in three-space." *J. Diff. Geom.* 16 (1981), 559–76.

[3] A. Bennett, "Surface Model." Available at http://www.sciencemuseum.org.uk/objects /mathematics/1996-544.aspx (February 2013).

[4] *Dehn Twist.* Available at http://en.wikipedia.org/wiki/Dehn_twist (February 2013).

[5] *Dirac's belt trick; Feynman's plate trick.* Available at http://en.wikipedia.org/wiki/Plate_trick (February 2013).

[6] J. Hass and J. Hughes, "Immersions of Surfaces in 3-Manifolds." *Topology* 24(1) (1985), 97–112.

[7] M. W. Hirsch, "Immersions of manifolds." *Trans. Am. Math. Soc.* 93 (1959), 242–76.

[8] R. Kusner, "Comparison Surfaces for the Willmore problem." *Pacific J. Math.* 138(2) (1989), 317–45.

[9] H. B. Lawson, "Complete minimal surfaces in S^3." *Ann. Math.* 92 (1970), 335–74.

[10] U. Pinkall, "Regular Homotopy Classes of Immersed Surfaces." *Topology* 24(4) (1985), 421–34.

[11] H. Samelson, "Orientability of hypersurfaces in R^n." *Proc. Am. Math. Soc.* 22 (1969), 301–02.

[12] C. H. Séquin, "Twisted Prismatic Klein Bottles." *Am. Math. Mon.* 87(4) (1980), 269–77.

[13] C. H. Séquin, "Tori Story," *Bridges Conf. Proc.,* The Bridges Organization, Winfield, KS, 2011, 121–30. Available at http://archieve.bridgesmathart.org/2011/bridges2011-121.html (August 2014).

[14] C. H. Séquin, "Torus Immersions and Transformations." UC Berkeley Tech. Rep. EECS-2011-83 July 2011. Available at http://www.eecs.berkeley.edu/Pubs/TechRpts/2011/EECS-20ll-83.pdf (April 2013).

[15] C. H.Séquin, "From Möbius Bands to Klein-Knottles," *Bridges Conf. Proc.,* The Bridges Organization, Winfield, KS, 2012, 93–102. Available at http://archieve.bridgesmathart.org/2012/bridges2012-93.html (August 2014).

[16] C. H. Séquin, "Topological tori as abstract art." *J. Math. Arts* 6(4) (2012), 191–209.

[17] C. H. Séquin, "Regular homotopies of low-genus nonorientable surfaces—Revised and extended version." UC Berkeley Tech. Rep. EECS-2013-21, March 2013. Available at http://www.eecs.berkeley.edu/Pubs/TechRpts/2013/EECS-2013-21.pdf (April 2013).

[18] C. Stoll, "ACME Klein Bottles." Available at http://www.kleinbottle.com/index.htm (February 2013).

Adventures in Mathematical Knitting

SARAH-MARIE BELCASTRO

I have known how to knit since elementary school, but I can't quite remember when I first started knitting mathematical objects. At the latest, it was during my first year of graduate school. I knitted a lot that year because I never got enough sleep and needed to keep myself awake during class. During the fall term I made a sweater for my dad, finishing the seams right after my last final, and in the spring I completed a sweater for my mom. Also that spring, during topology class, I knitted a Klein bottle, a mathematical surface that is infinitely thin but formed in such a way that its inside is contiguous with its outside (Figure 1). I finished the object during a lecture. It was imperfect, but I was excited, and at the end of class I threw it to the professor so he could have a look.

Over the years I've knitted many Klein bottles, as well as other mathematical objects, and have continually improved my designs. When I began knitting mathematical objects, I was not aware of any earlier such work. But people have been expressing mathematics through knitting for a long time. The oldest known knitted mathematical surfaces were created by Scottish chemistry professor Alexander Crum Brown. In 1971, Miles Reid of the University of Warwick published a paper on knitting surfaces. In the mid-1990s, a technique for knitting Möbius bands from Reid's paper was reproduced and spread via the then-new Internet. (Nonmathematician knitters also created patterns for Möbius bands; one, designed to be worn as a scarf, was created by Elizabeth Zimmerman in 1989.) Reid's pattern made its way to me somehow, and it became the inspiration for a new design for the Klein bottle. Math knitting has caught on a bit more since then, and many new patterns are available. Some of these are included in two volumes I coedited with Carolyn Yackel: *Making Mathematics with Needlework* (2007) and *Crafting by Concepts* (2011).

FIGURE 1. At left is an illustration of a Klein bottle, an infinitely thin, contiguous mathematical surface that intersects itself in three dimensions. At right is a knitted Klein bottle made in 2013. It is configured to mimic the illustration. Instead of a fixed self-intersection, the knitted object has a hole; it can thus be manipulated to demonstrate aspects of the surface. There are stripes on the surface that indicate Möbius bands knitted into the Klein bottle. See also color images. (All knitted objects made by sarah-marie belcastro. All photographs by Austin Green. All illustrations by Scientific American.)

You might wonder why one would want to knit mathematical objects. One reason is that the finished objects make good teaching aids; a knitted object is flexible and can be physically manipulated, unlike beautiful and mathematically perfect computer graphics. And the process itself offers insights: In creating an object anew, not following someone else's pattern, there is deep understanding to be gained. To craft a physical instantiation of an abstraction, one must understand the abstraction's structure well enough to decide which properties to highlight. Such decisions are a crucial part of the design process, but for the specifics to make sense, we must first consider knitting geometrically.

Knitting as Geometry

In a discrete geometric model for knitted fabric, plain knit stitches form a rectangular grid, or mesh, with one stitch sitting inside each rectangle (Figure 2). We shape knitted fabric primarily by using increases and

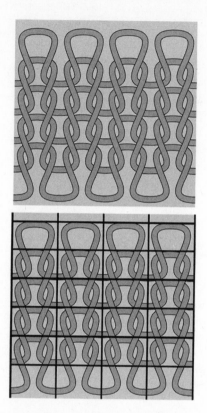

FIGURE 2. Rows of plain knit stitches have a regular, gridlike appearance (top). On the bottom, a rectangular grid, or mesh, is imposed on the stitches to show a discrete geometric model for knitted fabric. One stitch sits inside each rectangle; the mesh acts like a two-dimensional coordinate system. See also color images.

decreases. True to their names, increases add to the number of stitches in a row, and decreases lessen the number of stitches in a row. Both processes create mathematical curvature in the knitted fabric. Figure 3 shows an increase (angles and stitch sizes are not to scale). Interestingly, making a decrease has the same effect on the mesh; a decrease looks like an increase if the fabric is held upside down.

Rows and columns of stitches draw the eye along paths around the surface. Think of a sphere knitted from pole to pole: The rows and columns of stitches mimic latitude and longitude lines.

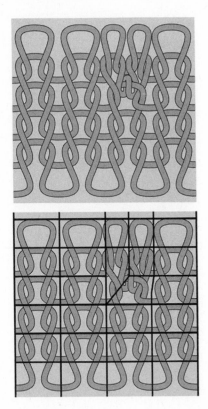

FIGURE 3. To add a stitch to a row, knitters use an increase (top), which can be represented in the polygonal mesh by two rectangles growing out of one (bottom). The lower edges of the two new rectangles are geometrically identified, although the upper edges are separate; thus the rectangles overlap, and an extra polygon is added to the mesh. The addition creates mathematical curvature in the knitted fabric. Angles and stitch sizes are not shown to scale. In reality, every stitch would be the same size, except for the lower halves of the two stitches in the increase, which would be compressed. Decreasing the number of stitches has the same effect; a decrease looks like an increase if the fabric is held upside down. See also color images.

All knitting is the generation of global structure via choices made in local stitch creation. A set of stitches appears to create a coordinate system (a grid in two-dimensional space). Because shaped fabric is not mathematically flat, however, any such system is only consistent locally, for that small patch of stitches.

FIGURE 4. Rows and columns of stitches draw the eye along paths around this sphere like latitude and longitude lines. Because shaped fabric is not mathematically flat, any coordinate system is only consistent locally.

An object that has consistent coordinates locally, but perhaps not globally, is in mathematical terms a *manifold*. Manifolds have a dimensional restriction: Every patch on a manifold must have the same number of coordinate dimensions. A pullover sweater represents a manifold—unless it has sewn-on pockets, because where a pocket joins the sweater, there are three different coordinate directions (up-down, left-right onto the pocket, and left-right under the pocket). A great deal of mathematical research concerns manifolds.

Most (but not all) knitted mathematical objects represent manifolds. In particular, although knitted fabric is of course three-dimensional, it represents two-dimensional things in the same way that paper represents an ideal 2-D sheet. That means that knitting can stand in for 2-D objects—or for the boundaries ("skins") of 3-D objects. Some mathematical objects are 2-D but cannot be represented in three dimensions without either intersecting themselves or having holes added. (These are objects that can be immersed, but not embedded, in real 3-space.) Klein bottles are in this class of objects.

The Case of the Klein Bottle

As mentioned earlier, a Klein bottle is an abstract, infinitely thin mathematical surface, formed in such a way that its inside is contiguous with its outside. That is, if it were thickened enough to have an inner skin separate from its outer skin, you could run a finger along the surface from any point on the outside to the corresponding point on the inside. Figure 5 shows a line drawing of a Klein bottle in three dimensions,

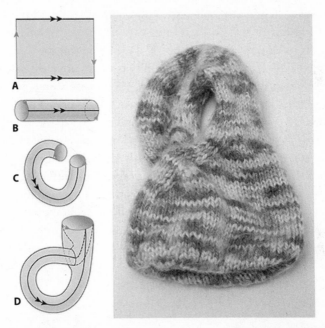

FIGURE 5. A Klein bottle can be represented in two dimensions as a marked rectangle (A, at left). Bending and stretching the rectangle through additional dimensions (B, C) allows us to glue the markings together, creating a three-dimensional model (D). My early Klein bottles were made with the knitting needles parallel to the single arrows and the rounds of knitting formed in the direction of the double arrows. The very first of these (right) was made on the fly from acrylic yarn. I knit a bulbous cylinder, then created a self-intersection in it by passing live stitches through the side of the cylinder, knitting a bit more, and grafting the cylinder's ends together. Because I did not add a twist before grafting the ends, the coordinates suggested by the rows and columns of stitches do not match the coordinates of the rectangle (A). See also color images.

along with a true 2-D representation as a marked rectangle; we bend and stretch the rectangle through additional dimensions to match up the arrows and glue them together. As the 3-D representation shows, the surface appears to pass through itself. At this place there must be a self-intersection or a hole.

In the terms of the 2-D representation, my first Klein bottle was made with the knitting needles parallel to the single arrows in the rectangle, and the rounds of knitting formed in the direction of the double arrows (Figure 5). It had a lot of flaws, but at the time, I only noticed one of them: It was immobile, stuck in one configuration in 3-space, because I had made it with a self-intersection. To address this problem, I promptly made a crappy Klein bottle with a hole (Figure 6).

The shortcomings of my early Klein bottles fell into two categories: aesthetic and mathematical. Aesthetics first: These things were *ugly!* One reason is that I used terrible materials. I have always preferred to use circular knitting needles, which are composed of two pointed tips connected by a thin, flexible cable. Back in graduate school, I often used hacked-together needles, shortening the cables of needles from the 1960s by melting the plastic (which promptly fell apart) or constructing my own circular needles from wire and packing tape (which caught on the yarn incessantly). These efforts to make needles that were both short enough and flexible enough for the projects resulted in a frustrating knitting experience and less than lovely results. Later I

FIGURE 6. My second knit Klein bottle was made using some unlabeled yarn I bought at a yard sale. I knit a cylinder with a slit in it and passed one end of the cylinder through that slit. Then I grafted the two ends of the cylinder together. Although this design was an improvement on the first, the coordinates suggested by the rows and columns of stitches still did not match the expression of the Klein bottle I wanted to create.

was able to get bamboo needles with hollow cables, which can be successfully modified. And I have learned how to use standard knitting paraphernalia to achieve the results I desire.

It wasn't just the needles, though—the yarn was a problem, too. That first Klein bottle was made from respectable but low-grade acrylic yarn. The second was made from a yarn of unknown provenance, which I found at a garage sale. It was awful to work with and had a sickening plastic-plus-fiberglass feel. These days, I tend to use high-quality rather than random yarns, and I choose the color, type of dyeing, and texture of the yarn with the intended mathematics in mind. Finally, in designing the objects, I improvised without much thought for details, so the first Klein bottle is misshapen, and the second has a weird vertical line on it (hidden in the figure).

Now for the mathematical issues. The approach I used in both cases was to knit something like a cylinder, pass one end of the cylinder through its side, and then graft the two ends of the cylinder together. One can build in a side slit (as I did with the second Klein bottle), or create a self-intersection by passing live stitches through the side of the cylinder (as I did with the first Klein bottle). Either way, using this method, one basically creates the Klein bottle from an identified rectangle, with the rows parallel to one pair of sides of the rectangle (single arrows in Figure 5) and the columns of stitches parallel to the other pair of sides of the rectangle (double arrows in Figure 5). It is very easy to graft the ends of the cylinder incorrectly. Unless the knitter introduces a half-twist before grafting, the four corners of the rectangle will not meet. Without the twist, the result is still a Klein bottle, but the coordinates suggested by the rows and columns of stitches do not match the coordinates of the underlying rectangle, and the geometry of the finished object is not as desired. A knitted cylinder, however, resists twisting—thus the difficulty in this step.

I eventually saw a solution to this problem: Create the twist intrinsically rather than adding it at the end. I accomplished this by starting with Miles Reid's Möbius band construction. This method effectively places the needles parallel to the double arrows shown in Figure 5, and knitting proceeds up both of the single arrows at once, each row wrapping twice across the rectangle. Because the twist is already present in the knitted fabric, the desired joining of the underlying rectangle happens automatically, with coordinates that match the rows and columns

FIGURE 7. This tiny Klein bottle (it's about 4 inches long) reflects my improved design—and better choice of yarn. Here I created the twist intrinsically rather than adding it at the end, by starting with Miles Reid's Möbius band construction. This design effectively places the needles parallel to the double arrows shown in Figure 5, and knitting proceeds up both of the single arrows at once, each row wrapping twice across the rectangle. Because the twisting is already present in the knitted fabric, the desired identification of the underlying rectangle happens automatically, with coordinates that match the rows and columns of stitches. See also color image.

of stitches. Skeletal instructions for knitting a Klein bottle in this fashion are available at my mathematical knitting website, and a photograph of one is shown in Figure 7.

That solution was the product of years of thought. Once I realized how flawed my first Klein bottles were, I wanted to do better, to create a design that was more faithful to the mathematics. Thus the question arose: What does it mean for a knitted object to be mathematically faithful?

Getting the Math Right

The way an object is constructed, in any art or craft, highlights some of the object's properties and obscures others. Modeling mathematical objects is no different: It requires that we make choices as to which mathematical aspects of the object are most important. When it's possible to do so, I knit objects so that a particular set of properties is intrinsic to the construction.

Most of the objects I make have both topological and geometric aspects. That is, these objects have an overall shape that is preserved when the objects are bent or stretched, and they have a specific form and structure in space. Sometimes the topology takes precedence and sometimes the geometry; this difference dictates whether and where curvature is placed on the object. Often I knit surfaces. These are 2-D and mainly smooth, which means that the constructions should have no seams, visible or otherwise. An ideal model of a surface should have no edges or bound stitches. In reality, yarn must have two ends. Similarly, a knitted piece may be made from more than one piece of yarn, but the transition from one ball to another should be invisible. These characteristics are part of a topological model for knitting, which I described in a 2009 paper in the *Journal of Mathematics and the Arts*.

Another consideration in constructing knitted mathematical objects is surface texture. For example, an abstract Klein bottle's inside adjoins its outside, so the texture of a knitted Klein bottle should be the same on the two physical sides; there should be no way of telling which real side is being looked at, and no way of identifying a transition between different parts of the knitted piece. For some other objects, the two sides are mathematically distinct and so their textures should be distinguishable.

There's more to consider—one can also indicate mathematics using a design knitted into the object. As a simple example, one can use stripes to indicate Möbius bands within a Klein bottle.

Most knitted-in designs are mathematically challenging because of the discretization problem: A smooth line or patch of color drawn on a piece of paper or electronically must be knitted as a sequence of discrete stitches. This harkens back to the mesh shown in Figures 3 and 4. Computer scientists who work on visualization of 3-D objects have developed algorithms for imposing a mesh on an idealized object. A finer mesh gives a smoother look, and in fact the use of very fine meshes is what produces realistic computer-generated imagery. In knitting, creating a finer mesh requires both a thinner yarn and substantially more time to complete the project. A great application of meshes to knitting appears in a 2012 SIGGRAPH paper, in which Cem Yuksel, Jonathan M. Kaldor, Doug L. James, and Steve Marschner explain how they use a mostly rectangular mesh to produce highly realistic virtual knitted garments.

Discretizing a pattern is not as simple as imposing a mesh. Knitted stitches are not square in proportion; they are close to 5:6 in aspect ratio. This fact has to be taken into consideration during design; if it is not, the finished object will be elongated. I usually make a swatch of fabric in the yarn I'm planning to use, measure to see what the stitch aspect ratio will be and then create graph paper with appropriately proportioned rectangles. Next one must determine the overall shape—where there will be increases and decreases, short rows, and other construction elements—and restrict the graph paper appropriately. The challenge comes in accurately placing designs or colors on that shape. Here are some examples.

First, consider the torus. A torus is a surface that is essentially a hollow doughnut shape—it is formed by rotating a circle in 3-D space about an axis. A torus knot is a closed path that can be drawn without crossings on the torus, so that when the rest of the surface is removed, the path is knotted. On a "flat" torus (similar to the "flat" Klein bottle drawn as a rectangle in Figure 5), a torus knot can be drawn as a straight path with a constant slope. Converting this to knitting requires changing the slope of the path with the curvature of the torus in a consistent way. Luckily, there is a knitted torus construction that minimizes the changes needed. Knitted torus knots are shown in Figure 10.

FIGURE 8. This Möbius band, knitted from Dream in Color veil-dyed yarns, is made with the complete graph K_6 knitted into it. Because of the small size of the object (it measures about 6 inches across), the corresponding coordinate mesh is coarse. Thus the edges in the graph are blurrier than those shown in the larger band in Figure 11. Each stitch has to be knit in a single color of yarn (the minimal mesh size is one stitch), and the surface texture dictates that adjacent stitches interlock colors. See also color image.

FIGURE 9. A torus is a mathematical surface with a hollow doughnut shape. I knitted the complete graph K_7 into the one shown above, using nine colors of Reynolds Saucy yarn. Designing a pattern so that multiple line segments are shown on a curved surface, such as that of a torus, is not easy. To distinguish between the line segments emanating from a dot, each segment had to have a different slope. Those slopes had to be modified in a manner consistent with the curvature of the torus in order to maintain the 7-fold symmetry. Added to this, the fact that the dots were of nonzero height and width made slope calculations even more difficult. See also color image.

FIGURE 10. These tori have trefoil knots knitted into them; the yarn is two colors of Southwest Trading Company Twize. The trefoil knot has two chiralities, so each torus knot is a mirror image of the other in structure as well as color. On a two-dimensional representation of a torus, a torus knot can be drawn as a straight path with a constant slope. Converting this to knitting requires changing the slope of the path in a consistent way as it traverses the curvature of the torus.

A further challenge is knitting a line or curve into an object with a bumpy surface texture. Each stitch has to be knit in a single color of yarn (because one stitch is the minimal mesh size), and the surface texture dictates that adjacent stitches interlock colors. Thus, to show a clear line, the line or curve has to be at least two stitches wide.

FIGURE 11. The Möbius band above, knitted from hand-spun wool/silk yarn, is designed to be worn as a scarf. Like the band in Figure 8, this one has the graph K_6 knitted into it. The graph edges are more clearly defined here because the object is larger and thus the mesh contains a greater number of stitches. See also color image.

Compare the graph drawn on both Möbius bands in Figures 8 and 11; one has a finer mesh than the other.

Knitting multiple line segments into a curved surface is also quite difficult. In another project, I knitted the complete graph K_7 into a torus, pictured in Figure 9. This project required the use of 7-fold symmetry. To make it possible to distinguish between the line segments emanating from a dot, each had to have different slopes, and those slopes had to be modified with the construction of the torus. Additionally, the dots were of nonzero height and width, which threw off slope calculations further.

The Design Process

So, what exactly does the design process for a mathematical object entail? Here is how I proceed. After deciding on an object to model, I articulate my mathematical goals (in practice, I often do this unconsciously). The chosen goals impose knitting constraints. This step gives me a frame in which to create the overall knitting construction for the large-scale structure of the object. Then I must consider the object's fine structure. Are there particular aspects of the mathematics that I can emphasize with color or surface design? Are particular textures needed? While solving the resulting discretization problem, I usually produce a pattern I can follow—my memory is terrible, and I would otherwise lose the work.

A recent mathematical creation can serve as a case study. A diagram in Allen Hatcher's *Algebraic Topology* had caught my eye, and I thought it would look fantastic knitted. The object shown is an equilateral Y extruded to be a three-finned thing with one end rotated by 1/3 and glued to the other end. Although, unlike a Möbius band, the object is not a manifold, it is a generalization of a Möbius band. So I thought I could use a similar construction—if only I could devise a way to knit outward from the central circle (the center of the Y). I wanted the knitted object to be created from a single strand of yarn because the mathematical object has a single edge. Thus, I had to create a way to use a single strand of yarn to produce three interlocking sets of free stitches. (Ordinary knitting has only two sets of stitches, upper and lower, per strand of yarn.) Once I had solved that problem—and it took me a while—I decided to use a texture that would look the same from all viewpoints, so that the central circle would be less visible. For my first attempt at the object, I decided to keep things simple and add no more requirements. The result is shown in Figure 12. After my first attempt was done, I took one look at it and realized that it resembled a cowl. I

FIGURE 12. The object above was inspired by a diagram in *Algebraic Topology*, a textbook by Allen Hatcher. It is an equilateral Y extruded across an interval, with one end rotated by 1/3 and glued to the other end. It was challenging to create this object from a single strand of yarn because I had to find a way to use that strand to produce three interlocking sets of free stitches—and ordinary knitting has only two sets (upper and lower) per strand of yarn. This iteration is wired to better show the structure.

FIGURE 13. This object is mathematically identical to that shown in Figure 12, but it is unwired and sized for wearing as a cowl.

resized the next version to produce a garment. A wearable mathematical object is a rare, and welcome, practical result.

Although I have worked on various knitting projects, I'm still not finished fiddling with designs for the Klein bottle—and it's been about 20 years since I began. I have been asked to adapt my construction into a wearable hat. It's one among many mathematical knitting challenges I look forward to completing.

Bibliography

belcastro, s.-m. The Home of Mathematical Knitting. http://www.toroidalsnark.net /mathknit.html.

belcastro, s.-m. 2009. Every topological surface can be knit: A proof. *Journal of Mathematics and the Arts* 3:2, 67–83.

belcastro, s.-m., and C. Yackel, eds. 2007. *Making Mathematics with Needlework: Ten Papers and Ten Projects*. Natick, MA: AK Peters.

belcastro, s.-m., and C. Yackel, eds. 2011. *Crafting by Concepts: Fiber Arts and Mathematics*. Natick, MA: AK Peters.

Dayne, B. 2003. Geek chic. *Interweave Knits*, Fall, 68–71 and 118.

Doyle, W. P. 2011. Past Professors: Alexander Crum Brown (1838–1922). University of Edinburgh School of Chemistry website. http://www.chem.ed.ac.uk/about/professors /crum-brown.html.

Hatcher, A. 2002. *Algebraic Topology*. Cambridge, UK: Cambridge University Press. Also available at http://www.math.cornell.edu/~hatcher/AT/ATpage.html.

Reid, M. O. 1971. The knitting of surfaces. *Eureka–The Journal of the Archimedeans* 34:21–26.

Yuksel, C., J. M. Kaldor, D. L. James, and S. Marschner. 2012. Stitch meshes for modeling knitted clothing with yarn-level detail. *ACM Transactions on Graphics* (Proceedings of SIGGRAPH 2012) 31:3. Available with video at http://www.cemyuksel.com/research /stitchmeshes/.

Zimmerman, E. 1989. *Knitting Around.* Pittsville, WI: Schoolhouse Press.

The Mathematics of Fountain Design: A Multiple-Centers Activity

Marshall Gordon

Introduction

Teachers of mathematics recognize the difficulty of reaching every student when the range of student abilities puts a considerable strain on the classroom discussion and time. In response to the problem, students are grouped so that those with greater mathematical aptitude help those who have difficulties. Although this approach is to be appreciated, it tends to mean that the more able students have less opportunity to explore further their own initiatives in mathematics, while those who have more difficulties find themselves on the receiving end with little opportunity to be in the role of enriching the mathematics experience for everyone, including themselves.

A "multiple-centers" approach is designed to overcome these problems. In this variation of differentiated instruction, the teacher introduces an area of investigation, say quadratic equations, by raising a particular problem, and after students—alone, together, and with the teacher's assistance—come to make sense of that initial focus problem, the students and teacher establish other problems that students get the chance to choose from as extensions of their quadratic equation investigations—in effect, creating multiple centers. After some time, the teacher brings the multiple centers of interest together for individual and group presentations, which may or may not be exclusively mathematical; some presentations may be more artistic, historical, or technological that are associated with or inspired by the initial focus problem. In this way, each student has the opportunity to gain the appreciation of their classmates inasmuch as everyone has something to offer as a consequence of their further multiple-centers inquiries.

Establishing the Focus Problem and
Its Ideational Offspring

The teacher's introduction of the quadratic equation by a discussion of the parameters of a parabolic arc determined by the path of a projectile over time given an initial velocity and height above the ground that is shaped by the effect of Earth's gravity sets the stage for the focus problem. Students see the resultant equation, $y = -(g/2)t^2 + vt + k$ as the height reached given the effect of gravity, an initial velocity, and an initial height.

The teacher then raises the question: is what we know sufficient to represent the parabolic trajectories in terms of associating heights with distances from the initial release point, such as the parabolic arcs in fountains, inasmuch as all that has been considered at present is the height of the projectile as a function of time, not distance?

To promote this investigation, there are many visual opportunities where students can appreciate fountain designs. For example, McGinley Plaza and Discovery Park Fountain at Purdue University in West Lafayette, Indiana, were inspired by the Fibonacci sequence (www.purdue .edu/dp/dptour/fibonacci.pdf). Additionally, Helaman Ferguson's Fibonacci Fountain at the Maryland Science and Technology Center in Bowie, Maryland, connects the Fibonacci sequence and the golden ratio. Ferguson's sculpture is based on the function $f(x) = (1 + \sqrt{5}/2)^{1/x}$ with $x = 0$ locating the point where the sculpture has its maximum height because this point would be where the water would theoretically shoot to infinity, and other values of x are chosen along the length in terms of successive Fibonacci ratios, determining the respective heights the water reaches. Yet, regardless of the mathematical or technological sophistication, from the common drinking fountains to the extraordinary fountains that surprise and delight us in special settings, we recognize the familiar shape. The water leaving the fountain surface projects upward, reaches a maximum height, and then returns, all the while following the path of a parabolic trajectory (Figures 1 and 2).

Student small-group discussions soon uncovered that the new context required rethinking their basic understanding of the parameters of the quadratic equation. A quadratic equation that would represent

FIGURE 1. A fountain in preparation.

FIGURE 2. Variations on an elegant curve. See also color image.

the parabolic paths of water would have to include the angle associated with the entry of the water into the air and would also have to be able to determine the horizontal distance traveled if the water is sure to stay within the bounds of the fountain. The class effort to determine the equations with these parameters sets the focus for the common investigation. And the context of creating fountain designs provides many opportunities for students of varying mathematical, artistic, scientific, and technical interest and facility to work alone and in small groups. Moreover, the setting provides the opportunity to include conversations about STEM considerations, where science, technology, engineering, and mathematics (STEM) are brought together.

The Focus Investigation

As with engaging any complex problem, considerable effort is required to formulate the problem so that it is in workable form. Accordingly, this aspect required the most consistent teacher intervention. Whereas representing the height of a concave parabolic curve as a function of time can be represented quite readily, with the new context requiring including the angle the water would exit the surface and the horizontal distance traveled as critical variables in their quadratic equations, the degree of complexity was raised considerably. Some students thought the variable of time would have to be replaced so as to write an equation in two variables: horizontal distance and the associated height. Yet they also understood how essential the variable of time was—they would need to take it into account to represent both the effect of gravity and the distance the water would travel as a function of the initial-exit velocity.

Some students sought to simplify the complexity by creating a right triangle with an angle of elevation representing the exit angle without considering the effects of gravity, though having both the horizontal distance, x, and the vertical distance, vt. This was an excellent application of the habit of mind, "make the problem simpler," which would come in handy on other occasions (see left side of Figure 3). Other students sought to replace the horizontal distance, x, in terms of the new context, by $vt \cos D$, and the vertical distance h by $vt \sin D$ (see right side of Figure 3).

From those initial understandings, two quadratic equations began to take form. Drawing upon the earlier consideration of combining the gravitational effect in conjunction with the linear distance the projectile would travel for a given time t, some students combined the height the water would reach, represented as $(\tan A)x$, where A is the assumed exit

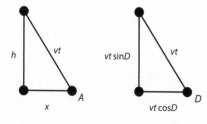

FIGURE 3. Getting an angle on the problem.

angle, in combination with the effect of gravity on the water as a function of time. Students working with this model began to consider how to eliminate t from their equation $h(x) = -(g/2)t^2 + (\tan A)x$. Their effort led them to realize that as $\cos A = x/vt$, they could solve for t in terms of x, resulting in $h(x) = -(g/2)/(v \cos A)^2 x^2 + (\tan A)x$. The students who had represented the horizontal and vertical distances in terms of the exit angle and time, so that $x(t) = vt \cos D$, adjusted the latter equation to include the effect of gravity as a function of time to $h(t) = vt \sin D - (g/2)$ $t^2 = (v \sin D)t - (g/2)t^2$, understanding that D was a constant, and that $x(t)$ would provide the horizontal distance as a function of time. The first equation allowed students to determine directly the height of the parabolic fountain arc as a function of the horizontal distance from the point where the water exited the fountain surface and the chosen exit angle. The second approach required two parametric equations, one to determine the height as a function of time, and the other to determine the distance associated with that height, also as a function of time. It made sense to everyone that zero would be the value of k in the standard quadratic equation because it expressed the idea that the parabola's trajectory began at the water's surface.

Now all students had equations that made sense to them, and yet had little or no idea what specific values for the exit angle, the horizontal distance determined by the entry point, or the time, would actually mean in terms of determining heights or more completely a fountain arc. Naturally, everyone tried various parameter values for their equations and graphed them, not unlike children playing in water, and came to the realization that both equations were valuable because they provided different perspectives and could answer different questions. Class and homework time were given to their associated investigations. The following section presents some of the investigations they pursued and the findings they made.

Multiple Centers of Investigation

1) Students inspired by water jets wanted to find the heights of other vertical water fountain jet streams associated with various velocities. (Informing the students that this curve was known as a "degenerate parabola" provided positive response.) Some Internet findings they made were as follows: King Fahd's

Fountain in Jeddah, Saudi Arabia, which reaches more than 300 m (1,024 ft), with water ejected at a speed of 390 km/h (242 mi/h), and the 130 m (426.5 ft) Jet d'Eau in Geneva, Switzerland, which has a speed of 210 km/h (130.4 mi/h). The Bellagio Hotel fountain in Las Vegas, New Mexico, claimed that 64,600 L (17,065 gal.) of water are in the air at the same time. Questions immediately began to surface: How much does the water in the air weigh? How much space does that much water take up? How long does the water stay in the air before returning to the surface? The more readily responded-to questions promoted a number of early experiences that essentially served to provide a sense of comfort and familiarity that made the consideration of more complex questions more available. It gave students who were less mathematically inclined more confidence that they could make headway and, it seemed, more resilience when things became less clear. The questions raised that were more complex served to create multiple centers of inquiry for a number of students.

2) Those students who had chosen the equation that included the tangent of the angle soon realized that they could not determine the height of a vertical jet of water with their formula as tan (90°) was undefined. However, those students with an equation in terms of time and the sine function were able to simplify their equation, for with sine (90°) = 1, they could consider vertical water jets using the equation $h(t) = vt - gt^2/2$. Some students drew upon the Internet data they had found to determine how long the water was in the air. Others realized that the maximum height, H, would occur along the axis of symmetry, $x = -b/2a$, from their earlier study of the general equation, $y = ax^2 + bx + c$. Solving for t in their equation, they found that $t = -v/-g = v/g$, and so, $H(v/g) = v^2/g - (g/2)$ $(v/g)^2 = v^2/2g$. Next, they used the heights of the fountain jets—data they had—and determined the velocities to reach those heights and checked their findings against those included in the Internet explorations. Knowing that $H = v^2/2g$ when $t = v/g$, they realized that they could determine how long the water that reached the maximum height was actually in the air if they would restate v in terms of t. Doing the substitution,

some students found that $t = 0.25(H^{1/2})$, and they were quite pleased with their algebraic facility; then other students realized that this was the time in one direction, and needed to be doubled, so that the maximum time, T, in the air would be $T = 0.5(H^{1/2})$.

3) The students, whose equation prevented them from considering vertical jets because the tangent is undefined at $90°$, put their energies to finding the angle that would create the greatest width of a fountain for a given velocity. A few decided that moving in increments toward $0°$ from $90°$ would uncover the angle by trial and error (a time-honored habit of mind) or at least a range within which the most valuable angle would appear. A few students recalled that the axis of symmetry was half the horizontal distance, so that doubling the horizontal distance would determine the point where the projected water would return to the surface. Using their equation, $h(x) = [-(g/2)/(v \cos A)^2]x^2 + (\tan A)x$, they determined the axis of symmetry, and after doubling and doing some trigonometric substitution, determined that the maximum distance, $X = 2(\tan A)(v \cos A)^2/g = 2v^2(\sin A)(\cos A)/g$. Now whatever the velocity, they sought to determine the angle A to create the maximum distance, X. The calculator was given considerable attention, and after graphing $y = (\sin x)(\cos x)$, the consensus was that it would occur when $A = 45°$, which confirmed their intuition. This result led to concluding that the maximum distance would be when $X = v^2/g$. They then determined the maximum height at that angle to be $H(45°) = v^2/4g$ and were pleased to discover that the ratio of maximum height to maximum width was 1:4.

4) Some students knew that their strengths were in drawing, and others understood that they could best enrich the fountain-design experience by doing historical research. So, while they participated in all the mathematics discussions, they were more observers listening to the questions and discussions, coming to understand and appreciate the mathematical findings rather than generating them. This work provided the opportunity for those students to make up equations after seeing others having done so and in the discussion process come to see the rationale

for their own equations. Those who did historical research made a presentation on early fountain designs that educated everyone about the engineering feat of the extraordinary tiered aqueducts that brought water from the surrounding hills to the citizens of ancient Rome. They also greatly appreciated how fountain designers in the seventeenth century used gravity to create fountains that had water jetting into the air without any mechanical (or of course electrical) technology. They also learned that fountains were not just for visual pleasure because most people collected their drinking water that way and used the runoff for washing. In comparison, fountains today may have pivoting nozzles controlled by computers that integrate lighting and music.

5) Student artistry and mathematics also enriched the shared mathematics experience to a delightful degree. In Figure 4, the student's parabolic water arc No. 1 is to the right of water arc No. 2, with associated equations $y = [-16/(65 \cos 75°)^2](x + 13)^2 + (\tan 75°)(x + 13)$ and $y = [-16/(65 \cos 75°)^2](x + 25)^2 + (\tan 75°)(x + 25)$, where the velocities of both arcs are 65 ft/s, and the water leaving the dolphins' mouths is at an angle of 75°. (Note: These equations are in feet per second because these students were more comfortable with this unit of measure.) The horizontal translation incorporated was appreciated by students, along with the fact that the dolphins were to be 36 ft (almost 11 m) tall, and the pool was to have a diameter of 100 ft (almost 30.5 m)—quite a visual spectacle. Interestingly, the student chose to locate the graphic origin of the two-dimensional form at the base on the vertical below the nose of dolphin No. 1. That the dolphins do not appear to be reflected across an imagined vertical is caused by the fact that the dolphin on the left is to be 10 ft (3 m) in front of the dolphin on the right. A number of students came up with a formula for the volume of a cylinder and the amount of water their fountain would hold. They were impressed by the weight of the contained water because a cubic foot weighed 62.4 lb (a cubic meter weighs 1,680 lb). (And though they expressed interest in building one, complications of time and space were prohibitive.)

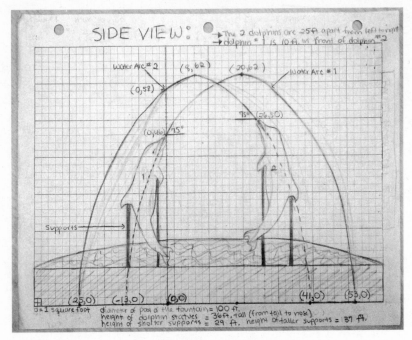

FIGURE 4. A student's solution.

Centers for Further Student Inquiry

Students made other investigations. Beyond those just presented, the following are offered in that spirit.

6) The physical apparatus to create a fountain using a hydraulic pump can be found at www.tryengineering.org/lessons/water fountain.pdf.

7) To determine flow rate assuming that the fountain water exits a cylindrical pipe, the volume that passes each second would be determined by the product of the velocity of the water and area of the pipe surface. Students need to experiment and be careful of units because pipe diameters are usually in centimeters for fountains.

8) The weight of the water of a parabolic arc could be approximated by considering the arc as if it were a cylinder with a diameter equal to the diameter of the pipe. Based on that form,

the volume could be determined, and as a result of multiplying by the density of water, an approximate weight of the water would be found.

9) To determine the length of the parabolic curve without using calculus, students can use their scale drawing and, laying a string along the curve, use that length adjusted to the original units. Some students may approximate half the length of the parabolic curve with the length of the hypotenuse of a right triangle, where the base of the triangle would be half the distance from where the water entered the air and returned to the pool, while the height would be at the vertex (Figure 5).

Students will surely recognize that the approximation suffers from the hypotenuse being a straight segment. So encouraging them to think about how they might adjust for the additional length in their approximation formula might yield their tinkering (another valuable habit of mind) with the coefficient of the height variable, which provides them with an experience that engineers have. That is, they could consider different values of k in the expression for approximating the entire curve length, $2\sqrt{(b/2)^2 + k(h^2)}$, $k > 1$ and compare those values with what would be found by measuring the parabolic arc lengths using the string. Civil engineer John Cresson Trautwine pointed out a century ago that setting $k = 4/3$ was "the approximate rule given by various [civil engineering] pocket-books" (1906, p. 192). Naturally, depending upon the magnitude of the curvature, the value is a better or worse fit. For example, he noted that if height = base/4, the approximation was about ½% too much, but if height = 10 × base, then the error went to about 15½% too much.

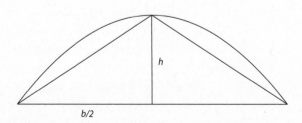

FIGURE 5. Seeking a good approximation.

10) Determining the quadratic equations representing water arcs at drinking fountains requires careful measurements. Students have to take into account that the point of entry is not at the same height as the end point in determining the equation representing the parabolic trajectory. Some students may discover that using the vertex can determine the angle measure inasmuch as $h = v^2(\sin A)^2/2g$ when $x = v^2(\sin A)(\cos A)/g$, and so $A = \tan^{-1}(2h/x)$.

11) Some students may be interested in determining how we actually get water coming out of sink faucets. They could learn how essential it is to make good decisions in locating reservoirs, and in laying miles of pipe, along with a realization of the practices and magnitude of the civil engineering involved. The economics involved could also be informative.

12) Others might be interested in comparing a fountain's parabolic trajectory on Earth with one on the moon, given that the latter's gravitational effect is approximately one-sixth that of the Earth's. Clearly, water exiting the fountain pool on the surface of the moon at the same velocity as on Earth will have a greater height and total flight time in the rarified air. Students will find that the total flight time will be around 2½ times as long as on Earth.

13) To appreciate the science of this technology, students might want to research laminar flow nozzles. These nozzles are constructed to remove turbulence so that the water arcs appear as if they are made of glass. There are many websites that demonstrate how to construct one.

14) Students artistically inclined might appreciate that it has been said that "architecture is frozen music" and create some forms, or series of jet heights, that "play out a tune." Others might want to create parabolic trajectories that suggest semicircular arcs, which would be in confluence with their circular fountain base—in effect serving as a rotation of the curve.

15) More can be done of course in terms of uncovering physical relationships, but some assistance in terms of the physics involved may be necessary. For example, it can be shown that the relationship between height in feet and pressure, P, in terms of psi, is $h = 2.31P$; and also that volumetric flow

rate $= 30d^2\sqrt{P}$, where d is the diameter of the water pipe and P is the pressure.

16) Those students who appreciate technology could consider radar and television "dishes" and uncover the rationale for their having parabolic shapes.

17) Some students whose interest is of a more abstract nature may wish to look into the mathematical origins of the parabola. The ancient Greeks who shared that interest were evidently compelled—after having determined that they could define a circle as a set of points at a given distance from a given point, they investigated what form would be created by a set of points at a given distance from a given point and a given line.

Conclusion

A multiple-centers inquiry is offered so that every mathematics student may find educational opportunities to enrich his or her own and others' awareness and understanding and can be appreciated for doing so. The openness of the inquiries tends to promote students working things out alone and together, and drawing upon mathematical habits of mind because they do not readily have algorithms or models to apply. Both aspects would suggest that the activity can invariably serve them well with regard to solving problems in many contexts. And if we want to promote a society where collaboration is valued, then educational institutions need to support that activity, despite the "time wasted" relative to the efficiency of a teacher lecture. It seems that the more such experiences students have for seeing how mathematics can provide insights into the world we inhabit and create, the more they are able to appreciate both the thinking of the ancient Pythagoreans and that of modern-day engineers, as well as their own.

Reference

Trautwine, J. (1906). *The Civil Engineer's Pocket-Book*, New York: Wiley.

Credit: "The Mathematics of Fountain Design: A Multiple-Centres Activity" by Marshall Gordon. *Teaching Mathematics and Its Applications* 32.1(2013): 19–27, by permission of Oxford University Press on behalf of The Institute of Mathematics and its Applications.

Food for (Mathematical) Thought

Penelope Dunham

Introduction

To paraphrase an old line, "Waiter, what's this potato chip doing in my calculus class?" The answer is, "It's a concrete model for a mathematical concept, and it generates extra interest for students as they learn that concept." In other words, the potato chip is an example of using food as a (tasty) manipulative to help students "touch it, feel it, learn it" in math class.

Moyer defines manipulative materials as "objects designed to represent explicitly and concretely mathematical ideas that are abstract" [6, p. 176]. The use of manipulatives in activities at all levels of mathematics instruction is endorsed by national organizations [7] and supported by a substantial body of research [9, 11]. Teachers also know that manipulative materials can add an element of fun to math instruction [6]. I have found that the "fun" element is particularly evident when manipulative materials involve food.

Although instruction with manipulatives is common in the lower grades, use of materials tends to level off as students get older and concepts become more abstract. Concrete materials, however, can still be effective at the undergraduate level, as I shall demonstrate with food-based activities involving exponential decay, saddle points, solids of revolution, and chi-square analysis. Moreover, explorations with food can be helpful even for abstract concepts such as strong induction.

One might ask, "Why teach math with food?" when there are a number of more typical manipulative materials available. My simple reply is that everything goes better with food (a fact well known to anyone who has organized a social event). Serving refreshments tends to increase attendance and interest, and that is true for math classes as well. In addition, I have found that food activities aid mathematical learning in ways that students remember long after the course ends.

In what follows, I organize the activities by course within three categories:

- using food as an effective example to illustrate a topic;
- using food as a concrete tool for exploration and concept development; and
- using food as a reward to motivate and spark interest in a topic.

In the first two categories, food plays a more direct role in learning through the hands-on activities that are the main focus of this article. The third category, however, also has value in the classroom; thus, I'll give a few examples in that area as well. (Instructors should check whether any students have food allergies or sensitivities. Doing so early in the semester allows time to find alternative foods for the activities.)

Using Food for Examples

In this section, I describe some food-based examples that demonstrate a particular topic in a concrete way. Typically, I use these activities soon after introducing a concept.

Statistics: Are colors uniformly distributed in bags of candy? That is the question students explore when I teach the chi-square goodness-of-fit test. I distribute small bags of three popular brands: Skittles®, M&Ms®, and Reese's Pieces®. Students then sort each sample by color and test the null hypothesis of equal proportions for the different colors of a given brand. Usually they do reject this hypothesis for Reese's Pieces® (three colors) but not for Skittles® (five colors). The M&Ms®, however, provide an added opportunity: a discussion of the effect of sample size on significance. With six colors, the initial samples are usually too small to reach significance; but, if we pool the class data, we find that colors of M&Ms®—unlike Skittles®—are not uniformly distributed.

Precalculus: Carrots provide a handy demonstration for a lesson on conic sections. Although not perfect, some carrots are close enough to a conical shape to show that cuts perpendicular to the axis form circles while cuts on a slant produce elliptical cross sections.

Calculus: A lesson on volumes of solids of revolution can be made concrete with some hard-boiled eggs and an egg slicer. Students see how an egg can be split into circular disks and then reassembled into its ovoid shape. The thin slices also demonstrate the notion of local linearity, in that the whole egg is curved but the edges of the thin disks look nearly straight.

Multivariable calculus: Pringles® potato snacks are a great way to illustrate saddle points. I distribute the potato chips and ask students to break them in each direction—a simple yet effective way to see that the "saddle" has a maximum in one direction and a minimum in another.

Using Food for Concept Development

The activities in this section go beyond a topic demonstration to involve more concept development than the examples in the first category. These hands-on explorations take a bit longer to complete and usually precede formal discussion of a topic.

Precalculus: A cup of M&Ms® provides a convenient simulation for exploring half-life or exponential decay because only one side of each piece is imprinted with an "m". (Note that instructors may substitute Skittles® (marked with an "S" on one side) for students who must avoid chocolate). Students count the candies, shake them in a cup, and pour them onto a paper plate. They remove any pieces without a letter showing, record the number of remaining candies, and repeat the process until all the candies are gone. Then students graph the number of pieces at each stage and discuss the shape. They try to model the output with a function and explain any deviations from the expected model. (An Internet search yields many versions of this exploration; for example, a student worksheet can be downloaded at [2].) This activity also works with pennies, but the students don't get to eat the data when they are finished!

Geometry and math for preservice teachers: Students often have incorrect notions about a relationship between area and perimeter or between volume and surface area. To address this misconception, I distribute 8.5×11 in. sheets of construction paper in two colors and ask my students to form two open cylinders by rolling one sheet with a landscape orientation and the other with a portrait format (Figure 1). We then discuss which cylinder to select if we wanted to fill it with M&Ms®. (Popcorn is a less expensive alternative.) Typically, most students think it makes no difference: the same amount of paper must result in the

FIGURE 1. Comparing volumes.

FIGURE 2. The surprising
result.

same volume. We then nest the taller cylinder inside the shorter one
and fill the tall one to the brim with candy. When we remove the inside
cylinder and see that the candy doesn't fill the outer one (Figure 2), a
lively discussion ensues. Next, we cut another sheet in half and tape
the pieces to form a tube half as high with twice the circumference to

surround the standing cylinder; then we repeat the activity by releasing the candy from the inner cylinder. Again, students are surprised by the result until we discuss the role that radius and height play in the volume formula.

Statistics: In a unit on confidence intervals for proportions, I use Gold-fish® crackers to help students understand the capture-recapture process for estimating the size (N) of a population. Using a large bowl as a "lake," we stock it with several packages of cheddar Goldfish®. Then we take an initial sample from the lake (the capture), find the number in the sample (M), "mark" them by replacing each (yellow) cheddar fish in the sample with a (brown) pretzel fish, and return the marked fish to the lake. After mixing thoroughly, we take a second sample of n fish (the recapture) and determine the number of marked fish (m) in the new sample. Using the proportion of marked fish in the sample, we get a point estimate for the proportion of marked fish in the lake from $M/N = m/n$ and use that to solve for N, the total population. We can also use our point estimate to construct a confidence interval for the proportion and find a range of values for the true population. For extensions of this activity and discussion questions, see [10, pp. 126–130].

Multivariable calculus: An article [5] about measuring the volume of an angel food cake pan gave me the idea for a week-long writing project in Calculus 3. After we have discussed centroids and volumes, I bring a pound cake and a tube pan (Figure 3) to class and ask the students how to determine the volume of the cake. Students can measure the pan, but its middle tube is slanted, and the pan's two-piece construction won't hold water for a direct measurement of volume. In addition, though the cake's straight sides conform to the pan's dimensions, the top of the cake is uneven because of rising. I then give each student a slice of cake and some grid paper. Their task is to trace the slice (Figure 4), estimate its area and centroid, and apply the first theorem of Pappus [3, p. 1031] to approximate the volume (cross-sectional area times the circumference of the centroid's path). They then document the methods they used in a short paper. Of course, they can eat the cake, too. (Another version of this problem is described in [4, p. 387]; it uses the shell method with heights along the parabolic cross section of a Bundt pan to find volume.)

Bridge course: The principle of complete induction is one of the more difficult concepts for students in a proof-writing course. There aren't

FIGURE 3. The cake pan.

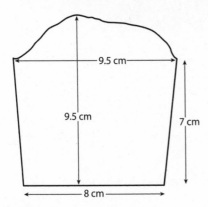

FIGURE 4. Cross section of a cake slice.

many easy examples that convey the necessity for the stronger induction hypothesis. I've found a tasty way, however, to explore the concept with Hershey® chocolate bars. The question is, "How many breaks (along grid lines) does it take to separate all the sections of a rectangular candy bar?" First, I distribute candy bars in different sizes and arrangements (e.g., 1×4, 3×4, and 3×6) and ask them to make a conjecture about the number of breaks required to separate n sections of a bar. For example, it takes three snaps to break a 1×4 bar into four pieces. Within a short time, they are convinced that we need $n - 1$ breaks for a bar with n sections; but how can we *prove* that? The base cases are easy: no breaks for one section; one break to separate two pieces. If we then assume that any bar with fewer than k segments can be split with $k - 1$ breaks, we are ready to use strong induction. Take a bar with k segments, and make one break along any of the grid lines to get two smaller bars, each with fewer than k segments (Figure 5). If,

FIGURE 5. Making smaller sections.

for example, our new bars have *r* and *s* pieces, respectively, we apply the induction hypothesis to get $1 + (r - 1) + (s - 1) = k - 1$ breaks. Voila!

Using Food to Motivate and Reward

The following examples don't use food for hands-on exploration, but they do create an environment in class that increases interest in a topic and thereby encourages learning.

Statistics: There are several opportunities in a statistics class to challenge students' intuition about the likelihood of certain events. In those situations, I often bet with my students about an outcome; if I lose, I'll bring them a treat for the next class. The birthday problem is a good example of such a scenario. How likely is it that we will find two people in our class with the same birthday? With a bet of brownies on the line, excitement builds as we record the birthdays on the board. I don't lose too often, but it's worth an occasional batch of brownies to see the class so engaged in the outcome. I use a variation when I introduce hypothesis testing by offering an Oreo® cookie to anyone who selects a black card from a deck of cards. The students don't know the deck is rigged (all red cards); they have assumed that it is fair. Attention is riveted on the package of cookies, as student after student fails to get a black card, until someone finally challenges the hypothesis of a fair deck. The cookies provide an incentive that makes the lesson memorable.

Calculus: After teaching the intermediate value theorem, I like to demonstrate a version of the bisection method with a game in which a student selects a word at random from a large collegiate dictionary. I then bet that I can find the word in fewer than 20 guesses. The idea, of course, is to narrow the choices by half each time (e.g., "Does your word come before or after "median"?). Naturally, food is the wager— usually, a pan of brownies. I have never had to pay that bet!

Calculus, abstract algebra, and other courses: Fundamental Theorem Day is a tradition that I learned from a former colleague, who said that major mathematical ideas deserve a party and a cake. After all, he would note, these results are *fundamental*. So, whenever we prove a big result in any class—such as the fundamental theorem of calculus in Calculus 1 or the fundamental theorem of arithmetic in our bridge course—I'll take a treat to class to celebrate the event and remind students that we've just proven something very important.

Conclusions

While writing this paper, I've learned other ways to use food to teach mathematics—ways that I look forward to incorporating in future courses. For example, recent articles have described how to find volumes of doughnuts and chocolate kisses [1] and a project in which calculus students find their own mathematical food connections [8]. One colleague's suggestion that particularly intrigued me involved using slices of an apple to teach level curves. The instructor made thin horizontal slices across the apple and had her students trace each slice in turn on a grid to see the shape of the apple emerge. I hope to use this activity the next time that I teach multivariable calculus.

I value the effect that hands-on explorations can have on mathematics learning. Although I've always used a variety of manipulative materials in my teaching, I believe that activities that involve food resonate in a special way with my students. Using food in my courses lets students "touch it, feel it, learn it" with tasty treats, have fun while learning, and . . . eat the data when they're finished!

Acknowledgments

Many thanks to the colleagues whose talks, articles, and conversations over the years have inspired me with examples of how to use food to teach math. Those that I remember include Gloria Dion (carrot conics), Margaret Dodson (hypothesis tests cookies), Bill Dunham (birthday problem brownies), Patti Locke (capturing goldfish), Mary Ann Matras (cake volume), John Meyer (fundamental theorem parties), Jasmine Ng (apple level curves), and Linda Taylor (exponential M&Ms®). I thought of the Pringles® by myself.

References

1. Bannister, N. A. 2012. A sweet way to investigate volume. *Mathematics Teacher.* 105(5): 400.

2. Clark, N. 2006. M&Ms and decay graphs. http://education.ti.com/calculators/down loads/US/Activities/Detail?id=6006. Accessed July 16, 2012.

3. Edwards, C. F., and D. E. Penney. 2002. *Multivariable Calculus, Sixth Edition.* Upper Saddle River, NJ: Prentice-Hall, Inc.

4. Finney, R. L., F. D. Demana, B. K. Waits, and D. Kennedy. 1999. *Calculus, Second Edition.* Reading, MA: Addison-Wesley Longman, Inc.

5. Matras, M. A. 1995. College mathematics corner: Angel food cake. *PCTM Newsletter.* 33(2): 25.

6. Moyer, P. S. 2001. Are we having fun yet? How teachers use manipulatives to teach mathematics. *Educational Studies in Mathematics.* 47(2): 175–97.

7. National Council of Teachers of Mathematics (NCTM). 2000. *Principles and Standards for School Mathematics.* Reston, VA: National Council of Teachers of Mathematics.

8. Petsu, E. 2011. A calculus food event. *Mathematics Teacher.* 104(8): 640.

9. Raphael, D., and M. Wahlstrom. 1989. The influence of manipulative materials in mathematics instruction. *Journal for Research in Mathematics Education.* 20: 173–90.

10. Scheaffer, R., M. Gnanadesikan, A. Watkins, and J. Witmer. 1996. *Activity-based Statistics.* New York: Springer-Verlag.

11. Sowell, E. J. 1989. Effects of manipulative materials in mathematics instruction. *Journal for Research in Mathematics Education.* 20: 498–505.

Wondering about Wonder in Mathematics

Dov Zazkis and Rina Zazkis

Prologue

We start with recreating a personal encounter.

A professor at a small university begins a lesson on fractals. He writes a simple-looking equation $Z_{n+1} = (Z_n)^2 + C$. He then describes an iterative process in which numbers (that correspond to points on the complex plane) are placed into the equation, yielding new numbers that are then placed back into the equation. At the same time, he quickly types an algorithm that tests the behavior of a grid of points. He finishes writing the code that accompanies his explanation, pauses for a second, says, "Here we go," and presses <ENTER>. The image in Figure 1 (known to mathematicians as the Mandelbrot set) appears on the screen, and the class pauses for a second in silent admiration.

The professor continues to explain the complexity of the pattern—that "zooming in" reveals progressively ever-finer self-symmetric recursive detail. As he demonstrates, a hand sticks up from the second row, shaking, indicating that its owner is eager to ask a question.

The professor points at the hand and says, "Yes, Michael." The hand retracts as Michael asks, "How does such a complicated pattern emerge from such a simple equation?" The professor freezes. He hesitates between an immensely complex and deep answer, having to do with the very fabric of complex numbers—something that is still being explored by research mathematicians—and the very simple, "It just does." He has no idea how to respond. Should he talk about the recursive process, or should he just ignore the question and extinguish Michael's interest in understanding the problem? After several seconds of awkward silence, a sarcastic comment emerges from the back of the room,

FIGURE 1. Mandelbrot set.

"Professor, what is the meaning of life?" The class laughs, and Michael's question is left unanswered, at least for that moment.

What Is Wonder in Mathematics?

The word "wonder," when used as a noun, is connected with admiration, awe, and surprise. When used as a verb, wonder is connected to interest and curiosity. This interest and curiosity lead to exploration and experimentation—what Hawkins (2000) referred to as "explorative inquiry"—and hopefully explanation and understanding. The noun and verb form are not separate; both can feed one another. Sensing wonder (noun) at a particular experience (that is, meeting it with surprise or admiration) may lead a person to wonder (verb) about its causes (that is, to question or to explore). On the other hand, wondering (verb) about a particular experience may lead to revelations that result in wonder (noun, that is, surprise or admiration). This distinction between the noun and verb form of wonder is also mirrored in the work of Fisher (1998), who explored how the wonder (noun) associated with the phenomenon of rainbows catalyzed wonder (verb), which led to the explanation of this phenomenon. This chapter's center of balance lies, delightfully, within Fisher's distinction. Though wonder and curiosity are deeply interconnected, Opdal (2001) draws an important distinction between these two epistemic notions. "Curiosity is a motive that can move a person to do all kinds of research, but within an accepted framework. . . . Wonder, on the other hand, is not a motive, but an experience or state of mind signifying that something that so far has been taken for granted is incomplete or mistaken" (p. 342).

Wonder—as described for example by Egan (1997) and Egan and Gajdamaschko (2003) as a cognitive tool—initially relates to fascination with extreme, strange, or surprising phenomena. Recognizing this tool, the teacher can highlight almost any object of study as "an object of wonder" (Egan and Gajdamaschko 2003, p. 89) by bringing out "the strange and wonderful in what seems routine or taken for granted" (Egan 1997, p. 219).

Wonder in mathematics reaches beyond a state of mind or fascination. Anne Watson, in Sinclair and Watson (2001), relating to teachers' appeal to students' awe with regard to patterns of nature or unexpected relationships, such as the equation $e^{ip} + 1 = 0$, describes this eloquently:

> I had a growing disaffection with this pedestrian approach to awe and wonder in mathematics, as if there were common sites for expressing awe, like scenic viewpoints seen from a tourist bus, whose position can be recorded on the curriculum as one passes by, en route for something else. Spontaneous appreciation of beauty and elegance in mathematics was not, for me, engendered by occasional gasps at nice results, nor by passing appeals to natural or constructed phenomena such as the patterns in sunflowers or the mathematics of tiling.
>
> (p. 39)

So, if not "spontaneous appreciation of beauty and elegance" of patterns and nice results, then what is "wonder-full" in mathematics? Michael's question in the beginning of this chapter goes to the heart of what wonder in mathematical contexts entails. The wonder in mathematics stems from the questions, "why?" and "how?" Unlike the fine arts, where just appreciating the awe of the end result—a painting, a dance, a poem—is the intention, mathematics places emphasis on process over product. When a mathematician reads a research article, he or she is often searching for new techniques. The beauty lies in the process of the solution, not in its end result. Standing back and appreciating a mathematical pattern or phenomenon could be the first step. Looking at an aesthetically pleasing mathematical object or one that acts in counterintuitive or interesting ways is not enough. But it can serve as a catalyst for cultivating wonder. This cultivation is in accord with Hadzigeorgiou's (2001, 2013) view of the role of wonder in science education.

Sinclair and Watson (2001) distinguished between two kinds of wonder: *wonder at* and *wonder why,* which correspond to the noun/verb distinction mentioned above. Sinclair and Watson pointed out that while wondering why and wondering how (the verb form) tends to lose its momentum once a satisfactory explanation emerges, wondering at (the noun form) can be a continual motivating force. As such, we can still find results or phenomena astounding even if we have a detailed explanation that accounts for them.

Our claim is that wonder—as related to mathematics—is the force that pushes us to move from an initial moment of aesthetic delight through the experience of creating intelligibility and understanding. As Hadzigeorgiou (2001) suggested, "wonder, in fact, gives things their meaning and reveals their significance" (p. 65).

In essence, we are exploring in our discussion how one can create what Harel (2013) termed an *intellectual need*—the need to understand why and/or how. We specifically focus on the following two of Harel's categories of intellectual need: the need for causality (why) and the need for certainty (how).

In the preceding example, Michael was filled with wonder (noun) at the sight of a Mandelbrot set that catalyzed his wondering (verb) about the deep connections between numbers on the complex plane that created the image. Even though an explanation of the Mandelbrot set is beyond the scope of this work, it is important to mention that the reciprocal relation between the two types of wonder, which Michael experienced, is the driving force behind advancement in mathematics. Of course, this is not true of mathematics exclusively. As Egan (1997) claimed, "Without the initial wonder, it is hard to see how more systematic theoretical inquiry can get fruitfully under way" (p. 97). Although this comment related to theoretical inquiry by scientists, specifically to Darwin's wonder at the variety of species in the Galapagos Islands, here we focus on a no less important kind of wonder—the wonder that can be created in students to drive their need to explore mathematics.

Types of Wonder in Mathematics

Many problem contexts in mathematics can create wonder. To narrow our discussion, we limit ourselves to several particularly telling examples of situations that, we feel, can elicit wonder. The main overarching theme in these examples is that of surprise: Michael's wonder, in our opening example, was caused not only by the beauty of the fractal image but also by the surprising fact that a simple equation can generate an infinitely complex drawing.

There are many opportunities for surprise in learning mathematics, but Adhami (2007) suggested that "Pedagogically we are focusing on students being 'taken aback' by a situation, hopefully causing them to 'look again' and spurring them to further effort to resolve an anomaly" (p. 34). Movshovitz-Hadar (1988) argued that "mathematics, at all levels is a boxful of surprises" (p. 34). In particular, she suggested that "all school theorems, except possibly a very small number of them, possess a built-in surprise, and [that] by exploiting this surprise potential their learning can become an exciting experience of intellectual enterprise to the students" (p. 34). We extend her claim by noting that surprise can be elicited not only in theorems but also by a variety of different mathematical situations.

Among potential mathematical surprises, we focus on the following themes:

- perceived "magic,"
- counterintuitive results,
- variation on a known result or procedure, and
- paradoxes.

Each of these things provides an avenue for creating a student's intellectual need to know and understand particular phenomena, to wonder why, to engage in exploratory inquiry, and to seek explanations. To highlight our approach to wonder in mathematics classrooms, we restrained ourselves to examples that require no more than a middle school mathematics background.

Perceived Magic

A clear example that engages almost everyone, regardless of age and education, is magic. Yesterday's Houdinis and today's David Copperfields and Criss Angels blow everyone's minds with their sophisticated tricks, which we often refer to as "magic." While enjoying the performance, there is an inevitable sense of wonder, the desire to uncover—and often to speculate about—how the trick "works." However, what appears as magical is often an issue of optical illusion and selective attention.

Is there magic in mathematics? Our claim is that by engaging students in what initially appears magical to them, we develop a sense of wonder and that this sense of wonder provides opportunities for them to wonder about a wide variety of mathematical situations. The question is, where are such situations found?

A Popular Introduction to Algebra

"I can guess your number" games are often used as a motivation for algebra. They start with: Think of a number, add 3 to it, subtract your number, add 1 to the result . . . and (magic!) I know that you got 4. In this oversimplified version, it is clear what happens to the number "in your mind," all the games of this kind eventually either subtract the original number after several manipulations or divide by it. Nevertheless, a teacher's ability to discover the number in a student's mind may invoke wonder and a desire to explore and understand the trick.

A more advanced form of the same idea is found in different versions of "How many times a week you want to do something," like eat in a restaurant, drink beer, or anything else that comes to your mind. In this kind of puzzle you enter the number of your choice, perform several arithmetic operations, and get a number that shows your age. Consider for example the following instructions:

- How many times a week do you like to eat chocolate? (chose a number 1 to 9, even if your "real answer" is higher than 9 or lower than zero).
- Multiply by 2.
- Add 5.
- Multiply by 50.
- Already had your birthday this year? Then add 1,762.
- Haven't had your birthday yet? Then add 1,761.
- Subtract the year you were born.

Surprise, surprise! In the 3-digit result, the first digit is the number you entered and the last 2 digits are your age. Wouldn't it be much simpler to subtract the year in which you were born from the current year? Nevertheless, this task is much more inviting than solving all even-numbered exercises on a given page, and an algebraic resolution is a worthwhile exercise. (Note: If you cared to go through the steps and didn't get your age, it is likely because this version of "chocolate math" was good for 2012. For this to "work" in other years, only the numbers 1,761 and 1,762 should be changed.)

MAGICAL MIND READER

In the above examples, it is clear that some mathematical manipulations determine the result, even when the specific nature of those can be complex. In what follows we present a less common and therefore a more "magical" example. We introduce a Web-based "Mind Reader," which can be found at http://www.flashpsychic.com/. One of the opening screens is shown in Figure 2.

So let us follow the instructions. Choose some number, say 31, subtract the sum of its digits, 4 (3 + 1 = 4) and get 27. Focus on the symbol next to 27, and—magic!—the symbol appears on the screen.

And there is an invitation to try again. We strongly recommend that the reader try again, connect to this website, and try again, and again.

So, did the Mind Reader read your mind? Let us try again:

We chose 84 this time, subtract 12 (12 = 8 + 4) and focus on the symbol next to 72 (72 = 84 − 12). Miraculously, or as expected, the symbol appears on the screen.

We have used this activity several times with both elementary school and university students. It is not uncommon for members of both groups to try to cover the webcams on their computers or face away from the screen, as if the Mind Reader was determining what number was in their head using some elaborate eye-tracking mechanism. Obviously, these actions do not prevent the Mind Reader from working. However, these reactions serve both to illustrate some rudimentary theory testing—"Is this website tapping into the webcam?"—and to demonstrate students' need to understand how this "Mind Reader" works, which is catalyzed by their curiosity.

The Flash Mind Reader

99 ♋	79 ❋	59 ❋	39 ♋	19 ♑
98 ☺	78 ♉	58 ♉	38 ♉	18 ❋
97 ♒	77 ☐	57 ♎	37 ♏	17 ♋
96 ♌	76 ♌	56 ☾	36 ❋	16 ♐
95 ☐	75 ▦	55 ♎	35 ♐	15 ☺
94 ♎	74 ♋	54 ❋	34 ☾	14 ✿
93 ♌	73 ☐	53 ♑	33 ♉	13 ♍
92 ♉	72 ❋	52 ☐	32 ◯	12 ▦
91 ♉	71 ✿	51 ❋	31 ♏	11 ♌
90 ♋	70 ▦	50 ♉	30 ☾	10 ♉
89 ♓	69 ❋	49 ☐	29 ☐	9 ❋
88 ☾	68 ▦	48 ♉	28 ♒	8 ♌
87 ♎	67 ☾	47 ♉	27 ❋	7 ☾
86 ♐	66 ♓	46 ☾	26 ♍	6 ♋
85 ☺	65 ☐	45 ❋	25 ❀	5 ♐
84 ❀	64 ✿	44 ♒	24 ♉	4 ♍
83 ☾	63 ❋	43 ♌	23 ☾	3 ♏
82 ❂	62 ♋	42 ♓	22 ♎	2 ☾
81 ❋	61 ♐	41 ♌	21 ♓	1 ♌
80 ☾	60 ♎	40 ♒	20 ♌	0 ♏

Choose any two digit number, add together both digits and then subtract the total from your original number.*

When you have the final number look it up on the chart and find the relevant symbol. Concentrate on the symbol and when you have it clearly in your mind click on the crystal ball and it will show you the symbol you are thinking of...

FIGURE 2. The Mind Reader opening screen.

In what follows, we will "spoil the magic," that is, explain how the magical effect is achieved. Readers interested in remaining under the spell of the magic are asked to skip to the next section.

The "Mind Reader" is cleverly designed so that different symbols appear next to numbers each time the game is played. (Compare for example the top line of symbols in Figure 4 to those in Figure 2—they are all different.) This design helps hide the fact that some symbols are strategically placed. You are encouraged to go back and pick different initial numbers in the above games. For example, if you try to start with the previously chosen number, 31, and apply it to the second game, you would end up at the symbol next to 27. It is the same snowflake that is placed next to 72, the number we got when we chose 84 as our initial starting point. For a reader interested in a hint, we recommend looking at all the other numbers that are listed next to the same symbol as 72 and 27. These numbers are 9, 18, 27, 36, 45, 54, 63, 72, and 81. What do all these numbers have in common and why? The reader is asked again either to pause to wonder, or to skip to the next section.

Try again!

FIGURE 3. The Mind Reader presents the number that has been calculated.

When the division of two whole numbers results in a whole number, we say that one number is divisible by another. In some cases, it is possible to determine divisibility without performing division, by considering the digits of the number. Such considerations are referred to as divisibility tests or divisibility rules. For example, the best known rule for divisibility by 2 considers the number's last digit: if a number ends in 0, 2, 4, 6, or 8, then it is divisible by 2. A less obvious rule deals with divisibility by 9: a number is divisible by 9 if the sum of its digits is divisible by 9.

Though this rule is often mentioned, practiced, and occasionally proved in middle school, its extensions are seldom mentioned. In fact, a number and the sum of its digits have the same remainder in division by 9 (*). To exemplify,

- $85 \div 9 = 9\text{Rem}4$, $8 + 5 = 13$, and also $13 \div 9 = 1\text{Rem}4$
- $64 \div 9 = 7\text{Rem}1$, $6 + 4 = 10$ and $10 \div 9 = 1\text{Rem}1$.

The divisibility rule for 9 is a special case of this statement: if the sum of the digits of a number has a remainder of zero in division by 9, then

The Flash Mind Reader

99 ♋	79 ☿	59 ♍	39 ♏	19 ♉
98 ❀	78 ♒	58 ♎	38 ☾	18 ❀
97 ♑	77 ☿	57 ♒	37 ♃	17 ♋
96 ❀	76 ♌	56 ♒	36 ❀	16 ☺
95 ♋	75 ♐	55 ☺	35 ☾	15 ☼
94 ♏	74 ☾	54 ❀	34 ☐	14 ☽
93 ♃	73 ♑	53 ♒	33 ♑	13 ♌
92 ♐	72 ❀	52 ♑	32 ♌	12 ♑
91 ☾	71 ♎	51 ♑	31 ♃	11 ❀
90 ♉	70 ☐	50 ♎	30 ♌	10 ♐
89 ☾	69 ♍	49 ♑	29 ☾	9 ❀
88 ♒	68 ♑	48 ♍	28 ☐	8 ❀
87 ☺	67 ☺	47 ♐	27 ❀	7 ♓
86 ☐	66 ♋	46 ❀	26 ♒	6 ♋
85 ☐	65 ♏	45 ❀	25 ♃	5 ♉
84 ☼	64 ❀	44 ♃	24 ♎	4 ♑
83 ♑	63 ❀	43 ♍	23 ♒	3 ♏
82 ♒	62 ♋	42 ☺	22 ☺	2 ♍
81 ❀	61 ☺	41 ♑	21 ♌	1 ♌
80 ☼	60 ☿	40 ♑	20 ❀	0 ♋

Choose any two digit number, add together both digits and then subtract the total from your original number.*

When you have the final number look it up on the chart and find the relevant symbol. Concentrate on the symbol and when you have it clearly in your mind click on the crystal ball and it will show you the symbol you are thinking of...

FIGURE 4. The Mind Reader, another opening screen.

the number itself has a remainder of zero. Furthermore, the result of subtraction of two numbers that have the same remainder in division by 9, is a number divisible by 9 (**). (Of course, this is true not only for the number 9, but only 9 is of interest here.) To exemplify,

- $85 - 13 = 72$ and $72 = 9 \times 8$, so 72 is divisible by 9
- $64 - 10 = 54$ and $54 = 9 \times 6$, so 54 is divisible by 9

By putting (*) and (**) together, we see that performing the arithmetic operation in the Mind Reader with any number of our choice—that is, subtracting the sum of the digits from the chosen number—we end up with a number divisible by 9. And all those have the same symbol in any iteration of the game. The clever design feature that changes the symbols every time the game is played masks this "divisible by 9" pattern.

To further mask the phenomena, the symbol that is placed next to the numbers divisible by 9 occasionally appears in other places as well. Also, 90 and 99 are marked differently. These two numbers are indeed divisible by 9, but an initial choice of a two digit number will not lead to

The Flash Mind Reader

Can't believe it? Click the button to try again.

Try again!

FIGURE 5. The Mind Reader presents the number that has been calculated.

either of these numbers. All these factors come together to contribute to hiding the "magic" behind the Mind Reader. Instead of trap doors and misdirection, a fascinating mathematical relationship is used (and perhaps still some misdirection).

Counterintuitive Results

The preceding discussion of a "magical" task demonstrated a type of wonder-inducing that required little pedagogical skill on the part of the instructor, at least when it comes to creating the wonder itself. This is not the case for all wonder-inducing tasks. In particular, motivating wonder through unexpected or counterintuitive results requires first skillfully setting up such expectations and/or intuitions. We believe that teachers have a substantial influence on students' expectations, both conscious and unconscious. Setting up expectations can be used as a pedagogical tool that sharpens, or even creates, students' ability to wonder. Without an appropriate setup, there is a danger that nothing will be found surprising.

Fisher (1998) argues that wonder occurs within breaks from the expected. He goes on to elaborate: "Awakening and wonder does not depend on awakening and then surprising expectation, but on the complete absence of expectation" (p. 21). If taken literally, which we do not believe was Fisher's intention, we disagree with this statement. Our functioning as human beings is deeply tied to our predictive expectation of the world; it is difficult to separate out everyday functioning from these expectations. If, however, this statement is viewed through a more figurative or metaphorical lens, we believe that it offers an important insight that we can use in our quest to wield wonder as a pedagogical tool, namely, that the break from the expectation from which wonder emerges is often most powerful when we are not overtly conscious of what our expectations are. In such situations, unconscious presuppositions must be brought to the forefront in order for them to fit a newfound experience.

To illuminate this discussion on setting up subconscious expectations, let us take for example a lesson on regular polyhedra, also known as Platonic solids (see Figure 6).

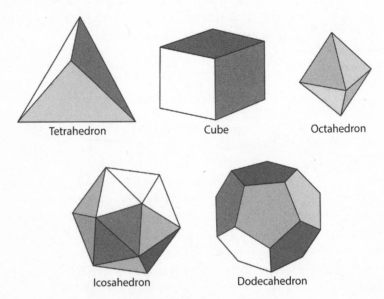

Tetrahedron   Cube   Octahedron

Icosahedron   Dodecahedron

Figure 6. Platonic solids.

PLATONIC SOLIDS

Setting expectation can begin with a discussion of regular polygons. There are equilateral triangles, squares (or regular quadrilaterals), regular pentagons, hexagons, heptagons, octagons, etc. How many different types of regular polygons are there? Students quickly, if not instantaneously, realize that they can always just add an additional side to get a new polygon. So there is a regular polygon with 12, 20, or 100 sides. There are infinitely many different ones. Once this idea is established, we have set up an expectation for the three-dimensional analog of regular polygons, namely, for regular polyhedra.

When students discover that building progressively more complicated polyhedra is not as simple as the two-dimensional case, they are genuinely surprised and intrigued. By facilitating students' discovery of this apparent discord we have set up wonder. The existence of only five Platonic solids at first appears counterintuitive because it is not in accord with the two-dimensional case. Had we instead started directly with the three-dimensional case, then the surprise, the wonder, and the desire to explain this result would have been much harder to come by. The explanation for why there are only a few Platonic solids concerns the angles of polygons that meet "at a corner." When equilateral triangles (which have all angles of 60 degrees) meet at a corner (we need at least three to create a corner), the sum of angles can be 180, 240, or 300 degrees, when a corner is created by 3, 4, or 5 triangles, respectively. The resulting polyhedra are tetrahedron, octahedron, and icosahedron, respectively. Note that an attempt to create a corner with six triangles "fails" because the sum of the angles is 360 degrees and they "lie flat" on a plane. Similarly, a corner can be made with three squares ($3 \times 90 = 270$ degrees) and 3 pentagons ($3 \times 108 = 324$ degrees), resulting in a cube and dodecahedron, respectively. All the other combinations cannot create a corner because they create a surface that is either flat (for example, 4 squares with 360 degrees sum of angles meeting at the corner) or is concave (for example, four pentagons with the sum of corner angles of 432, which is larger than 360). Here we attempted to provide the reader with the intuition behind the result, rather than with a rigorous argument. A complete and formal proof for the existence of exactly five Platonic solids can be found, for

example, at Wolfram Math World, http://mathworld.wolfram.com/
PlatonicSolid.html.

COUNTERINTUITIVE PROBABILITIES

Other examples of counterintuitive results come from expectations
that students set up themselves. A famous example is the Monty Hall
"Let's Make a Deal" game. The game introduces the player to three
doors, where there is a valuable prize (e.g., a red Ferrari) behind one
of the doors and something undesirable (e.g., a goat) behind two other
doors. Though we can imagine a situation in which a goat is more useful
than a Ferrari, it is assumed that winning the Ferrari is the desirable
outcome. The player chooses one of the doors. Regardless of his choice,
Monty selectively opens one of the other two doors to reveal an undesirable prize. The player is then offered an opportunity to change his
initial choice to the other still-closed door. So the player is presented
with a dilemma—to switch or to stick with his or her initial choice.

Every time we introduce this scenario in class—unless students have
previous experience with the solution—there appears to be an agreement that "this does not matter." The treasure is behind one of the two
still-closed doors, and therefore there is a 50% chance for success of
each decision.

However, the fact that there are two possible results does not mean
that there is a 50% chance of winning regardless of our choice. In a
lottery, there are also two possible outcomes—you either win or don't
win—and nobody thinks that there is a 50% chance of winning the lottery. In the "Let's Make a Deal" scenario, the initially surprising fact is
that by switching his initial choice, the player doubles her or his chances
of winning. Numerous articles have been written on this issue. Several
websites (e.g., http://www.grand-illusions.com/simulator/montysim
.htm) not only provide a detailed explanation but also include simulation of the game that can be carried out a large number of times. No
definite conclusion is possible after running 10 experiments. But after
100 or 1,000 times, experiments show that the probability of success
when switching is close to 2/3. This counterintuitive and surprising
result creates a sense of wonder and consequently the desire to understand the situation. The class discussions attempting to resolve the
player's dilemma are often loud and even emotionally charged.

We turn now to a less familiar, and as such maybe more "exemplary," example (Mason 2006). *If Mrs. Chance has two children, and one of them is a boy, what is the probability that the other one is also a boy?* Assuming that the probability of having a boy in any childbirth is ½ and that the gender of one child has no influence on the gender of another, it is reasonable to conclude that the answer is ½. However, this is incorrect—do you wonder why?

Let us consider all the possibilities of genders in a family of two children. It could be (1) boy, girl (2) boy, boy (3) girl, boy, or (4) girl, girl, where these four cases are equiprobable, that is, have the same probability of occurring under the assumptions listed above. However, we know that one of Mrs. Chance's children is a boy, so option (4) does not apply in her case. Focusing on cases 1, 2, and 3, we see that the other child is a boy only in one of the three equiprobable cases. As such, the probability that Mrs. Chance's other child is also a boy is ⅓—a result that most find surprising because it is not in accord with the initial intuition.

Even more surprising is the interpretation of results of medical testing. Suppose that you were diagnosed with a rare terminal disease X, for which there is no cure, and the testing is 99% accurate. What are the chances of you actually having the disease? The initial reaction is to assume that if the results of the testing are 99% accurate, then it is almost sure (that is, 99% sure) that you have the disease. However, rather than wondering what prayer your family will choose for your funeral, let us focus on numbers. Mathematics can actually entail some good news. If the rare disease has an occurrence of 1 in 10,000, then the probability of your dying from this illness is about 1%. That is, there is a 99% chance that you will live happily ever after this unfortunate diagnosis. This is because the 1% of people without the disease who get a false positive diagnosis is much larger than the 99% of people with the disease that get a positive diagnosis. This is an example of how wondering about numbers and learning to interpret them makes a difference, not only between success and failure on a mathematics test, but also between party and funeral preparations.

Variation on the Familiar

Remember column multiplication? Recall that the numbers had to be written in a certain form, one underneath another, and then, for some unexplained reason, some rows had to be moved to the left. That is what our Grade 2 teacher taught, Grade 5 teacher reinforced, and parents approved. Students still learn how to do this as if calculators, like remote controls, are chronically lost between couch cushions whenever they are needed. Because the same algorithm is taught to different generations of students, one gets an impression that that's the one and only way to do multiplication by hand. However, to some students' surprise, this is not the only way. One variation is known as "Russian peasant multiplication" or sometimes as "the Egyptian algorithm." It could be the case that the ancient Egyptians visited Russian peasants, or vice versa. However, it is more likely that in the precalculator era, great minds thought alike.

To introduce the algorithm, consider multiplying two numbers, for example, 114 and 23. We start by writing down 114 on the top of the left column and 23 on the top of the right one. We double numbers in the left column, and divide by 2 numbers in the right column, writing down the quotient and ignoring the remainder, until we reach 1.

114	23
228	11
456	5
912	2
1824	1

We highlight with a (⋆) the rows in which the remainder in division was ignored, that is, all the odd numbers. We then add the corresponding numbers from the left column.

114	23	⋆
228	11	⋆
456	5	⋆
912	2	
1824	1	⋆

That is, $114 + 228 + 456 + 1{,}824 = 2{,}622$, which is indeed $2{,}622 = 114 \times 23$.

Of course because of the commutativity of multiplication, it does not matter what's on the right and what's on the left.

23	114
46	57
92	28
184	14
368	7
736	3
1472	1

Similarly to the above, we double the numbers in the left column and divide by 2 numbers in the right column, ignoring the remainder.

23	114	
46	57	⋆
92	28	
184	14	
368	7	⋆
736	3	⋆
1472	1	⋆

We highlight with a (⋆) the odd numbers in the right column and add the corresponding numbers from the left column: $46 + 368 + 736 + 1,472 = 2,622$.

We often introduce this multiplication algorithm to prospective elementary schoolteachers. Our students not only learn to perform multiplication as Russian peasants or ancient Egyptians did, but they are also eager to understand how this works. That is, they wonder why it works the way it does. Surprisingly or not, they have never wondered about the mystery of the "conventional" algorithm.

To satisfy the curiosity of a reader not previously familiar with Russian Peasant multiplication, we provide a brief explanation. We note the addends in the first case: 114, 228, 456, and 1,824. However, these numbers, other than the beginning 114, are the results of doubling:

$$114 = 114 \times 1$$
$$228 = 114 \times 2$$
$$246 = 114 \times 4$$
$$1{,}854 = 114 \times 16$$

Therefore, their sum produces the desired product.

$$114 + 228 + 246 + 1{,}854 = 114 \times 1 + 114 \times 2 + 114 \times 4 + 114 \times 16$$
$$= 114 \times (1 + 2 + 4 + 16) = 114 \times 23.$$

Similarly, considering the second case,

$$46 = 23 \times 2$$
$$368 = 23 \times 16$$
$$736 = 23 \times 32$$
$$1{,}472 = 23 \times 64$$

Therefore,

$$46 + 368 + 736 + 1{,}472 = 23 \times 2 + 23 \times 16 + 23 \times 32 + 23 \times 64$$
$$= 23 \times (2 + 16 + 32 + 64) = 23 \times 114$$

Working with prospective teachers, we invite them to notice "something special" about the numbers used here, that is, 1, 2, 4, and 16 in the first case and 2, 16, 32, and 64 in the second case. Our lesson may proceed toward binary representation of numbers, but this is beyond the scope of our discussion here. What is important, though, is that the wonder about the unconventional algorithm can be satisfied and then turned, with a gentle pedagogical move, to wonder about the ordinary, that is, about the conventional algorithm. In our experience, this method results in better understanding of the conventional and a new appreciation of the distributive property and of the reason behind rules that were previously perceived as "strange."

Simpson's Paradox

A different source of wonder is found in scenarios that appear upon initial inspection to be paradoxical. These cause intense deliberations in mathematics classrooms, based upon some profound disagreement.

Consider for example the following scenario: at our university, the admission to the teacher certification program is rather competitive. Only a small number of applicants are admitted to the program; the (fictitious) admission numbers from a recent competition are presented below.

	Applied	Admitted	Admitted %
Men	2,400	720	30%
Women	7,300	1,640	22.5%

As shown, the percentage of women admitted to the program (22.5%) is much lower than the percentage of men (30%). What could be the reason for this? Are women less prepared and less qualified? Or is there a systematic bias against women that could be the basis for a discrimination lawsuit against the university? These questions may generate an interesting classroom discussion, especially in a class of students admitted to the program. However, let us examine the numbers more carefully. Below, the total numbers of applicants are separated into 3 groups: applicants for certification at elementary school level, applicants for secondary school humanities, and applicants for secondary school sciences and mathematics.

	Secondary Math/Science		Secondary Humanities		Elementary		Total	
Men	1,000	50%	1,000	16%	400	15%	2,400	30%
	500		160		60		720	
Women	300	60%	3,000	22%	4,000	20%	7,300	22.5%
	180		660		800		1,640	

The numbers are chosen to be close to a realistic situation: the admission for elementary education and secondary humanities is much more competitive than the admission to secondary sciences. By examining each category separately, we see that the percentages of female applicants that are admitted is indeed higher than the corresponding percentages of male applicants. However, focusing on the total, we see that overall a higher percentage of men are admitted.

How come? Students' first reaction is usually to recalculate the numbers. Once the correct calculation is confirmed, there is a sense of confusion, but also a sense of wonder. How come? What's the trick?

In fact, the presented numbers are a simplified analogy to the famous gender bias case related to the admission to graduate school at the

University of Berkeley in 1973. (See, for example, http://en.wikipedia
.org/wiki/Simpson%27s_paradox.) This is an illustrative example of
what is known in statistics as "Simpson's Paradox." That is, the relationship present in different groups is reversed when the groups are combined.

The reality is that in our scenario of admissions to the teacher certification program, similarly to the University of Berkeley case, women
seem to apply to more competitive program routes, such as elementary
education and secondary humanities. The number of male applicants to
the program routes with relatively high admission rates, secondary sciences, is much higher than the number of female applicants.

Well-known cases of Simpson's paradox relate to batting averages
(when Player A performs better than Player B in all terms, but when
the totals are combined Player B outperforms player A) and health-care
disparities. We exemplify the latter.

Suppose that a pharmaceutical company developed two different
drugs to treat X. (To add humor to the situation, X can be laziness
or sleepiness in a classroom.) To check the effectiveness of the drug,
it was administered to different groups of volunteers. The results are
presented below.

	First try		Second try	
Drug A	100 patients	90%	100 patients	82%
	90 successes		82 successes	
Drug B	30 patients	93.33%	300 patients	83.33%
	28 successes		250 successes	

Obviously, after administering the treatment for the first time, Drug
B showed better results. However, since Drug B was administered to a
much smaller group, the size of this group was increased for the second
administration. The results yet again favored Drug B.

	First try		Second try		Combined	
Drug A	100 patients	90%	100 patients	82%	200 patients	86%
	90 successes		82 successes		172 successes	
Drug B	30 patients	93.33%	300 patients	83.33%	330 patients	84.24%
	28 successes		250 successes		278 successes	

However, the combined results favor Drug A. The choice is not simple. The classroom discussion will remain inconclusive. The obvious solution is to deal with similar size groups in such experiments. However, this is a solution for the future drug administration, which applies neither for the current case, nor for cases—such as a pool of applicants to a certain department—where a group size is not controlled. The importance of these examples is that they can be presented in a way that creates a dissonance and, in that respect, encourages conversation, which ultimately leads to a deeper understanding of mathematics.

Conclusion

While the noun "wonder" (wonder at) is associated with admiration and awe, we have highlighted "wonder" as a verb (wonder why and wonder how), which is associated with interest, curiosity, surprise, and exploratory inquiry.

Wonder in mathematics is often expressed as

a. the desire to understand and/or prove an observed relationship and
b. the desire to test extensions, variations, and possible generalizations.

We argued that genuine surprise—when a result is both unexpected and unexplained—is the main catalyst for wonder in mathematics. We focused on how "wonder at" (unexpected/surprising results and relationships) can be cultivated into "wonder why." This takes care of (a). With respect to (b), we trust that extended experiences with "wonder why" will eventually result in further "wonder why." That is, understanding a relationship will result in a desire to explore it further, to test the scope of applicability.

For example, having explored divisibility by 9 in wondering about the magic of the Mind Reader, one may inquire and investigate other divisibility tests. Having seen several examples of Simpson's Paradox, one may vary several numbers and investigate under which conditions there is still a "paradoxical" relationship. Having explored the Russian peasant multiplication algorithm, one may become interested in investigating additional nonstandard algorithms.

Sinclair (2006) referred to this desire to test variations and extensions as "looking for more" (p. 53). She described her experience in solving a problem in geometry. Having examined a particular case, she described her pursuit as follows:

> . . . the possibility of a framing structure was emerging . . . I was anticipating that certain relationships would emerge in a whole family of shapes. As with reading a novel, I wanted to find out what would happen next, what theme would emerge from the sequence of ideas. (p. 51)

The "chase" that Sinclair described is familiar to those of us who enjoy mathematics. Adhami (2007) described this as "you stay puzzled for a while even after you resolved the issue and new questions arise spontaneously in your mind" (p. 34). We believe that with an appropriate pedagogical approach, such puzzlement can become contagious for students. According to Movshovitz-Hadar (1988), "It is the mathematics teacher's responsibility to recover the surprise embedded in each theorem and to convey it to the students" (p. 39). This responsibility applies, as we mentioned previously, not only to theorems but to a wide range of mathematical structures and relationships. However, presenting a surprise, and subsequently inducing wonder, depends not only on a teacher's skill but also on carefully chosen tasks. Whereas magic (such as in the Mind Reader) and novelty (such as in Russian peasant multiplication) are more obvious choices for sources of surprise, Adhami noted that a potential disagreement is one of the task components that helps capitalize on students' surprise in a classroom. Our examples of Simpson's Paradox and Mrs. Chance's children are sources of dissonance and significant disagreement among students that lead to fruitful discussions.

"Capacity to wonder is also an attitude towards experience" (Sinclair 2006, p. 130). We believe that presenting students with experiences that cause them to wonder not only fosters their desire to resolve and extend (or, in Piagetian terms, assimilate) these situations but also develops their ability to understand situations (algorithms, formulas, relationships) that have been perceived as mundane.

References

Adhami, M. (2007). "Cognitive and social perspectives on surprise." *Mathematics Teaching,* 200, 34–36.

Egan, K. (1997). *The Educated Mind: How Cognitive Tools Shape Our Understanding.* Chicago: University of Chicago Press.

Egan, K., and Gajdamaschko, N. (2003). "Some cognitive tools of literacy." In A. Kozulin (Ed.) *Vygotsky's Educational Theory in Cultural Context* (pp. 83–98). Cambridge, U.K., and New York: Cambridge University Press.

Fisher, P. (1998). *Wonder, the Rainbow and the Aesthetics of Rare Experiences.* Cambridge, MA: Harvard University Press.

Hadzigeorgiou, Y. (2001). "The role of wonder and 'romance' in early childhood science education." *International Journal of Early Years Education,* 9(1), 63–69.

Hadzigeorgiou, Y. (2013). "Reclaiming the value of wonder in science education."In E. Kieran, A. I. Cant, and G. Judson (Eds.) *Wonder-full Education: The Centrality of Wonder in Teaching and Learning across the Curriculum.* (pp. 40–65). Abingdon, U.K.: Routledge.

Harel, G. (2013). "Intellectual need." In K. Leatham (Ed.) *Vital Directions for Mathematics Education Research.* New York: Springer.

Hawkins, D. (2000). *The Roots of Literacy.* Boulder: University Press of Colorado.

Mason, J. (2006). "What makes an example exemplary: Pedagogical and didactical issues in appreciating multiplicative structures." In R. Zazkis and S. R. Campbell (Eds.) *Number Theory in Mathematics Education: Perspectives and Prospects* (pp. 41–68). Mahwah, NJ: Lawrence Erlbaum Associates.

Movshovitz-Hadar, N. (1988). "School mathematics theorems—An endless source of surprise. *For the Learning of Mathematics,* 8(3), 34–40.

Opdal, R. M. (2001). "Curiosity, wonder and education seen as perspective development." *Studies in Philosophy of Education,* 20, 331–34.

Sinclair, N. (2006). *Mathematics and Beauty: Aesthetic Approaches to Teaching Children.* New York: Teachers College Press.

Sinclair, N., and Watson, A. (2001). "Wonder, the rainbow and the aesthetics of rare experiences." *For the Learning of Mathematics,* 21(3), 39–42.

The Lesson of Grace in Teaching

Francis Edward Su

We know truth, not only by reason, but also by the heart.
—Blaise Pascal

I'm honored, but I'm also really humbled to be giving this talk to a room full of great teachers because I know that each of you has a rich and unique perspective on teaching.* I had to ask myself: Could I really tell you anything significant about teaching?

So I decided instead to talk about something else, something that at first may appear to have nothing to do with teaching, and yet it has everything to do with teaching.

I want to talk about the biggest life lesson that I have learned and that I continue to learn over and over again. It is deep and profound. It has changed the way I relate with people. It has reshaped my academic life. And it continually renovates the way I approach my students.

And perhaps it will help you frame your own thoughts about teaching. The beginning of that lesson is this:

Your accomplishments are *not* what make you a worthy human being.

It sounds easy for me to say, especially after having some measure of academic "success" and winning this teaching award. But 20 years ago,

---

* This text was originally a talk given on the occasion of receiving the 2013 Haimo Award for Distinguished Teaching from the Mathematical Association of America. After giving this talk, I had so many requests for the text that I posted it on my blog: http://mathyawp .blogspot.com. It was the hardest thing I ever had to write because it is deeply personal, truly me, and about my biggest life lesson . . . given at a conference in front of hundreds of people who, I'm sure, struggle with the same things that I do.

I was a struggling grad student, seeking validation for my mathematical talent but flailing in my research, seeking my identity in my work but discouraged enough to quit. My advisor had even said to me: "You don't have what it takes to be a successful mathematician."

It was my lowest point. Weak and weary, with my identity and my pride stripped away and my Ph.D. nearly out of reach, I realized then that my identity and self-worth could *not* rest on whether I succeeded or failed to get my Ph.D. So *if* I were to continue in mathematics, I could not do it for any acclaim that I might receive or for the trappings of what the academic world would call "success." I should only do it because math is beautiful, and I feel drawn to it. In my quiet moments, with no one watching, I still found math fun to think about. So I was convinced that it was my calling, despite the hurtful thing my advisor had said.

So did I quit? No. I just changed advisors.

This time, I chose differently. Persi Diaconis was an inspiring teacher. More than that, he had shown me a great kindness a couple of years before. The semester I took a class from him, my mother died and I needed an extension on my work. I'll never forget his response: "I'm really sorry about your mother. Let me take you to coffee."

I remember thinking, "I'm just some random student, and he's taking me to coffee?" But I really needed that talk. We pondered life and its burdens, and he shared some of his own journey. For me, in a challenging academic environment, with enormous family struggles, to connect with my professor on a deeper level was a great comfort. Yes, Persi was an inspiring teacher, but this simple act of kindness—of authentic humanness—gave me a greater capacity and motivation to learn from him because we had entered into authentic community with each other, as teacher and student, who were real people to each other.

So when the time came to change advisors, I decided to work with Persi, even though it meant completely starting over in a new area. Only in hindsight did I realize why I had gravitated to him. It's because he showed me grace.

Grace: Good things you didn't earn or deserve, but you're getting them anyway.

By taking me to coffee, he had shown me that he valued me as a human being, independent of my academic record. And having my worthiness separated from my performance gave me great freedom! I could truly enjoy learning again. Whether I succeeded or failed would not affect my worthiness as a human being. Because even if I failed, I knew: I am still worth having coffee with!

Knowing that my new advisor had grace for me meant that he could give me honest feedback on my dissertation work, even if it was hard to do, without completely destroying my identity. Because, as I was learning, my worthiness does *not* come from my accomplishments. I call this the lesson of grace.

The Lesson of Grace

- **Your accomplishments are *not* what make you a worthy human being.**
- **You learn this lesson when someone shows you grace: good things you didn't earn or deserve, but you're getting them anyway.**

I have to learn this lesson over and over again.

You can have worthiness apart from your performance.
You can have dignity independent of achievements.
Your identity does not have to be rooted in accomplishments.
You can be loved for who you are, not for what you've done—
somebody just has to show you grace.

You are worth having coffee with!

Now the academic world does not make it easy to learn this lesson, especially when so much of academic success depends on achievement: grades, degrees, publishing papers, getting tenure. And we are applauded for those achievements. We crave that applause! So it's tempting to be drawn into this trap of needing my achievements to justify me.

So even now, as I receive this award, I must hold fast to this lesson. I must not cling to this award too tightly. It does not *give* me dignity because if someone showed me grace, I'd realize I already *have* dignity.

Don't get me wrong. I'm not saying that achievement shouldn't be rewarded. There *is* a place for credentials in academia. We would not want to hear a talk by someone without credentials. We would not want to graduate students who didn't have skills. But achievement, in its rightful place, is *not* where we should derive our ultimate sense of identity and self-worth, and we need to have a healthy separation between achievement and worthiness.

If I can really believe this, then it gives me great freedom! I can do math simply because I enjoy it, not because I have to perform. I don't have to be "the best." I can stop being so hard on myself. I can have healthy ambition without competition: striving toward goals, without having to compare myself to other people. I can be happy for another person's success. I can be appropriately open and authentic. I don't have to fear showing weakness. Because my worthiness isn't earned, there's no need and no room for pretense. I can stop worrying about what others think of me if I believe the lesson of grace.

Grace seems simple, but it is a deep concept. Once you recognize it, you begin to see it everywhere. Some might recognize grace as a part of many of the world's great religions. That makes sense because at its core it's a theological concept, making a claim about who we are as human beings, and why. In my own religious view, I see Jesus as the ultimate giver and source of grace, endowing all human beings with worth and dignity that they don't have to earn. But whether you are religious or not, everyone can give, receive, and be drawn to grace, graceful actions, and its lessons because *grace gives people dignity they don't have to earn.*

What does this life lesson have to do with teaching? Well, if life is one gigantic learning experience, then you'd expect that any life lesson we learn would shape our teaching. But the lesson of grace has remarkable implications. Here are four ways I see that grace can shape our teaching. These go from easiest to hardest: giving grace to students, understanding grace in our teaching, communicating grace in the struggle, and sharing grace in our weakness.

Giving Grace to our Students

What does it mean to give grace to our students?

The first example is something we already all do. What do you do when you want to be nice to your students and you want to wake them up at 8:00 in the morning? Yes, you give them donuts! They didn't have to earn that. That's grace! (Except on evaluation day—then it's a bribe.)

Here's another way we show grace to our students—learning their names. By naming people, you give them *dignity*. Imagine the other possibility: suppose you only learn the names of the people who are getting A's or coming to office hours. That's not grace because it dignifies only the people who earn it.

Spend time with students outside class. That's grace: it's a good thing they didn't have to earn. As long as it's not just the best students you hang out with, then it's grace.

I have often given fun exam questions: students can earn some easy points just by sharing the most interesting thing learned in the class, or a question they'd like to pursue further. Or "Write a poem about a concept in this course." Or "Imagine you are writing a column for the newspaper: 'great ideas in math.' What would you put in it?" These are graceful questions. They really didn't have to earn those points, and they're having fun while doing it.

And of course, sharing the joy of mathematics is grace—and going off on tangents in class. Many of you know that I have a collection of "math fun facts." I have often started off calculus lectures with five-minute "math fun facts" that have nothing to do with calculus, just to get students excited about mathematics. This is a graceful action. Because going off-topic communicates something to students: that they can learn math just because it's cool, not because they have to "get through some material" that they'll be tested on.

There's a website where you can find my collection (search for "Math Fun Facts" to find it), and if you prefer a mobile version, there's an app for that!

Understanding Grace in our Teaching

If we fully understand the lesson of grace, then we'll understand this idea: since my performance doesn't define me, I don't have to be the center of attention in my classroom. I can do experimental things, and fail. I can get out of the way of my students. I can open up the classroom for things like inquiry-based learning. I don't have to be in control of everything. I don't have to worry about what people think of me.

I'll give a recent example of where I had to think about these things. A couple of years ago, a former student came to me with an idea. He was creating an online learning platform, and he wanted to pair up videos of my Real Analysis course with scrolling notes and social learning features, and I said: "That's interesting. . . . what would it involve?"

He said, "We would just have to record your class and put it on YouTube!" And I hesitated.

Then he said, "It would be cool if my class could try out the software and we could run some experiments." What he was suggesting sounded to me like a radical overhaul of the way I would teach my class. And it made me nervous. This is getting to be a bit much, I thought.

But upon reflection, I realized that the only reason I hesitated is because I was fearful of losing control, fearful of crazy Internet comments and what others would think of me. And I could extend him a great grace by helping him pursue his passion.

So I agreed to have the class taped. What's interesting is the unexpected grace that occurred as a result of the YouTube experiment. The students were excited about it. They loved the fact that they could watch the videos later. They didn't stop coming to class, as I had worried about. And to my surprise, I began to get grateful e-mail from people around the world. Many of them didn't have access to a university, were facing particular economic hardships, or learned best when they could pause and rewind lectures.

For them, the videos were a grace they didn't have to earn. At the beginning of the semester, I had thought I would just take down the videos at the end because I was so worried. But I never did because I realized that they are serving a needed function for the least fortunate in our global community and for people who learn differently.

Communicating Grace in the Struggle

I want to demonstrate to my students that their worthiness does not depend on the grades they earn in my class. Of course, I want to give my C students the same attention that my A students get. But if I am really honest with myself, I have to admit that I like talking to A students because they "get it." They already speak the same language.

But what credit is it to me as a teacher if I only affirm the students who already "get it"? It's easy to affirm the student who asks great questions in class, but I must be thoughtful about how I can affirm the questions from a struggling student, or the one who comes from a different cultural background, or the one whose educational system didn't provide him or her with the needed tools. How can I affirm these students?

I like to tell them that the struggle is the more interesting place to be—because a healthy confusion is where the real learning begins. Just as in life, the most meaningful lessons are learned when our afflictions and struggles are greatest.

But I want to be clear: I am not saying that extending grace is a recipe for helping my students feel good about themselves. I am saying that it will help them have a right understanding about themselves. So if my students know in their bones that I have given them a dignity that is independent of their performance, then I can have honest conversations with them about their performance. I can judge their work justly *and* graciously. In fact, failing a student can be done with grace, so that the student understands that his or her dignity has not been tarnished even though the work has been justly assessed—just as a parent can discipline her or his child if the child knows that the parent's love is unconditional. Grace is precisely what makes hard conversations possible, and productive, between people. But you have to extend the grace first.

I want the failing student to understand clearly that grades are just an assessment, not a sentence. I try to meet with every failing student in person, and I carefully articulate the distinction between their grade and their worthiness. I often give them this explicit word of encouragement: that although grades attempt to measure what you have learned, they do not measure your dignity as an individual.

Sharing Grace in our Weakness

I don't mind telling students that I almost didn't make it in graduate school. Because I understand that my worthiness is not in my accomplishments, I don't fear that people will think less of me. I know what it means to enter a program with a weaker background than my peers, to feel woefully underprepared, to feel misunderstood, to have family pressures that suddenly become paramount, to wonder if I was really cut out for this profession. So I know that weakness can be powerful when a former student shares:

> He gave me the single most important piece of advice I got before heading to graduate school, which greatly shaped how my mathematical career developed. It occurred when I asked him about

his graduate school days, which surprising as it may be, did not go very smoothly for him! He confessed to me that at one point he considered dropping out of Harvard! The lesson he learned was to pick an advisor you can . . . thrive with, even to the sacrifice of a particular subject or project. I took this advice to heart . . . and as a result I thrived in graduate school which has directly resulted in my early career success as well.

This is from a student who was not at the top of his class at Harvey Mudd, but he chose a graduate school where he could thrive; it led to an NSF postdoc, and he's just finishing that now.

I don't mind talking with students who are having serious family issues about losing both of my parents to terminal illnesses and telling them that it's okay to let academic work suffer—because as human beings, they aren't defined by their academic work.

I don't mind telling students with emotional issues that it's okay to see a counselor—because I've seen a counselor.

So with a struggling student, showing my weakness is extending and sharing grace. I am validating their worthiness in our shared struggles. *They don't have to perform well to earn my respect.*

And sometimes, showing weakness enables us to receive grace from our students. One of the nicest things a student ever said to me came when my father was dying of cancer, and I flew back and forth to Texas multiple times to tend to his care. There was a point in the semester when my class had had more lectures from other people than from me. It was surely disruptive for them to see a different professor every day. So I confessed to my class that I had two roles—as a son and as a teacher—and I felt that I was doing neither of those roles well. One of my students said to me gently, "Should I be terminally ill later in life, I would want my son to act as you have."

Ah, grace! From my student, who reminded me: I didn't need to be so hard on myself. *I didn't need to perform well to earn his respect.*

So This Is the Lesson of Grace

- **Your accomplishments are not what make you a worthy human being.**
- **You learn this lesson by receiving grace: good things you didn't earn or deserve, but you're getting them anyway.**

And this is my hope: that you can receive and give grace.

We are so trained by our accomplishment-driven culture to believe that our deeds are what make us worthy of honor or respect. To fight this belief, you have to surround yourself with grace givers, people who are good at it.

All the best teachers in my life have been grace givers. Think of that teacher whom you knew was busy but still made you feel like you were the most important person in the world. Think of those people whom you can be authentic with, those who, even if they know all the rotten things about you, would love you anyway—the ones around whom you feel you have no shame.

Sure, good instructional techniques are necessary for good teaching. But they are not sufficient. They are not the foundation. Grace-filled relationships with your students are the foundation for good teaching because grace gives you freedom to explore, freedom to fail, freedom to let students take control of their own learning, freedom to affirm the struggling student by your own weakness. Grace amplifies the teacher-student relationship to one of greater trust in which a student can thrive.

To teach is to create a space in which the community of truth is practiced.
                                                      —*Parker Palmer*

That community and space that Parker Palmer talks about does not form without grace.

I like to think that I'm a good teacher because I communicate well and I choose the best examples, and that when my former students think of my teaching, they think of these things. But that is accomplishment-driven thinking, isn't it? Instead, what students remember most often are those moments of grace.

Last year, at a Harvey Mudd graduation event, a math major named Simeon was invited to give a speech to parents about his college experience. I'd like to close my talk by sharing part of it, with his permission:

The one class that best embodies the essence of Harvey Mudd College was a class called Real Analysis.

In Real Analysis I learned to question the very definition of real numbers and everything I knew about mathematics. What do you mean I have to prove how to add two real numbers? Proof by common sense and elementary education were strictly prohibited.

Real Analysis was perhaps the hardest class I've taken, and my first experience of struggling in math. I wasn't getting the concepts as quickly as some of my peers, and I couldn't help feeling incompetent in math, a subject I had always felt confident in . . . the "gateway" to mathematics never felt so narrow and without space for an incompetent student like me.

Fortunately, there's more to the story. During that semester I was doing a book study with Professor Su outside of class, and I was uncomfortable. Sitting before me was a super smart incredible professor, and I felt really unworthy to be hanging out with him because I wasn't doing so well in his class, and I thought I might disappoint him once he got to know me personally. But at our last meeting, we were talking and he said, "I want students to understand that professors don't value students based on their academic performance" . . . to hear from my own professor, whom I really love and admire, at a time when I felt ashamed of my intelligence and thus unworthy of his friendship, that I wasn't just a student in a seat, not just a letter grade or a number on my transcript, but a valuable person who he wants to know on a personal level, was perhaps the most incredible moment of my college career. And that's the kind of place that Harvey Mudd was.

Yes, Simeon, you get it! You understand the transformative power of grace! My hope for all of us is that we can understand grace in all its forms and how it can transform our teaching.

And not only will grace inspire our students, it will inspire us. Just like my students, the moments I remember best from my own teaching are the grace-filled moments I have shared with my students and colleagues and former teachers, many of whom are here today. I want to thank them—because I didn't deserve those blessed moments. But they gave them to me anyway.

Generic Proving: Reflections
on Scope and Method

URI LERON AND ORIT ZASLAVSKY

A generic proof is, roughly, a proof carried out on a generic example. We introduce the term *generic proving* to denote any mathematical or educational activity surrounding a generic proof. The notions of generic example, generic proof, and proof by generic example have been discussed by a number of scholars (Mason and Pimm 1984; Balacheff 1988; Rowland 1998; Malek and Movshovitz-Hadar 2011). All acknowledge the role of proof not only in terms of validating the conclusion of a theorem but, just as importantly, as a means of gaining insights about why the theorem is true. In particular, we support and extend the argument made by Rowland (1998) that a generic proof does carry a substantial "proof power" and may in fact lie on the same continuum as the working mathematician's proof. In the same vein, we analyze possible ways that generic proof and proving may help in unpacking and making accessible to students at all levels *the main ideas* of a proof [1].

This article is organized as a reflection on three examples, or "mathematical case studies," which reveal increasingly more subtle facets of generic proving. The first mathematical case study is a simple and elementary theorem on numbers (also discussed in Rowland 1998). The second example, a decomposition theorem on permutations, is still elementary in the sense of not requiring subject-matter knowledge beyond high school mathematics but is more sophisticated in terms of the proof techniques required. The third example, Lagrange's theorem from elementary group theory, is more sophisticated both in terms of the proof techniques and the subject matter knowledge required. All the examples are introduced in a self-contained manner, and all the terminology is explained and exemplified.

In the second part of the article, we reflect in more depth on the mathematical case studies of the first part, in an attempt to explicate some of the general features of generic proofs. For example, in an attempt to characterize the mathematical content of generic proofs, we look for commonalities with professional mathematicians' proofs as they appear in research journals and in university-level textbooks and lectures. For another example, we ask—and try to give some partial answers—about the *scope* of generic proving: what kind of proofs can be more or less helpfully approached via a generic version?

The article has been written in the form of a thought experiment. It is, however, solidly based in the experience of the authors in running many workshops with students and in international conferences on exactly these examples and ideas. Several researchers have previously discussed the more theoretical aspects of generic proofs. This research, while relevant to the topic at hand, would take us away from our mathematical and pedagogical focus [2].

Mathematical Case Study 1: Counting the Factors of a Perfect Square

Theorem: *A natural number which is a perfect square (i.e., the square of another natural number) has an odd number of factors.*

For example, the number 16 ($= 4^2$) has 5 factors (namely: 1, 2, 4, 8, 16), and 25 ($= 5^2$) has 3 factors (namely: 1, 5, 25).

Generic Proof: Let us look at the perfect square 36 ($= 6^2$). We want to show that it has an odd number of factors. We list systematically all the factorizations of 36 as a product of two factors:

$$1 \times 36$$
$$2 \times 18$$
$$3 \times 12$$
$$4 \times 9$$
$$6 \times 6$$

All the factors of 36 appear in this list. (We could go on listing 9×4, 12×3, etc., but because multiplication is commutative, this would just repeat the previous factorizations and would not produce new factors.) Counting the factors, we see that the factors appearing in all the products, except the last, come in pairs and are all different, thus totaling to an even number. Since the last product, 6×6, contributes only one factor to the count, we get, in total, an odd number of factors. Specifically, we have $2 \times 4 + 1 = 9$ factors.

From this first simple example, we can already get an initial idea of what a generic proof is, and of some of its strengths and weaknesses. Obviously, our generic proof is not a complete proof, since the theorem has only been proved for the particular number 36. However, the number 36 was treated as *generic* in the sense that we did not make use of any of its specific properties except that it is a perfect square. In fact, all the important ideas of the general proof already appear in this generic proof, with the result that students could easily reproduce the proof for any other example. Indeed, they would most likely feel that they were carrying out *the same* proof. Thus, a generic proof serves as an easy introduction to the proof's *main ideas*.

Note the choice of 36 as our generic example. We felt that 36 could represent for the learner *any* perfect square, whereas 4, 16, 25, or even 169 ($= 13^2$) would have been too special to highlight the generalizability of the proof (e.g., they would have too few factorizations). In Rowland's (1998) words, it is "small enough to be accessible with mental arithmetic but with sufficient factors to be non- trivial" (p. 68). In Mason and Pimm's (1984) terms, it allows us "to see the general in the particular" (p. 277).

We partly concur with Movshovitz-Hadar's (1988) suggestion that a generic example should be "large enough to be considered a non-specific representative of the general case, yet small enough to serve as a concrete example" (p. 17). In general, however, "size" should be replaced by a measure of the *complexity* of the example. In the case considered here, complexity is measured by the number of factors, not the magnitude of the number; thus, 169 is less generic than 36, since the former is too special, having only 3 factors. In general, the example chosen should be "complex enough" to ensure that all the main ideas of the target proof will naturally surface in the context of the example.

Mathematical Case Study 2: Decomposing a Permutation into Cycles

Theorem: *Every permutation has a unique decomposition as a product of disjoint cycles.* (These terms will be explained as the proof unfolds.)

To highlight both the mathematical and educational aspects of generic proving, we present the theorem and its proof via a thought experiment of an idealized virtual classroom scenario [3]. In their previous lesson, the students in our scenario have already learned and practiced the definition of a permutation (a one-to-one mapping of the set $\{1, 2, \ldots, n\}$ onto itself) [4] and the 2-row notation for permutations [5]. They have also learned when two permutations are equal (when they are equal as functions) and the definition of multiplication for permutations (i.e., perform the two mappings in succession, the same as composition of functions).

1. *Teacher:* Let's look at an example of a permutation, say the permutation σ below, and see if we can find anything interesting about its structure—how it can be constructed from simpler permutations (similarly to how numbers are constructed from their prime factors).

$$\sigma = \begin{pmatrix} 1 & 2 & 3 & 4 & 5 & 6 & 7 \\ 6 & 1 & 2 & 4 & 7 & 3 & 5 \end{pmatrix}$$

For example, let's start at 1, and follow its path as we apply the permutation σ over and over again, thus: $\sigma: 1 \rightarrow 6 \rightarrow \ldots$

[The students work in teams, continuing what the teacher has started: $\sigma: 1 \rightarrow 6 \rightarrow 3 \rightarrow 2 \rightarrow 1$]

2. *Alpha:* It came back to 1! There is no point going on, since it will just repeat the same numbers.

3. *Teacher:* Right. This part of the permutation is called a *cycle*, and is written (1 6 3 2). It is a special kind of permutation, in which each letter in the cycle notation goes to the next one on the right, except the last one, which goes back to the first. (The letters that don't appear in this notation are understood to be mapped to themselves; for example, in this cycle, $5 \rightarrow 5$.) Note that the same cycle can also be written as (6 3 2 1), (3 2 1 6), or (2 1 6 3), since they are all equal as functions.

   Let's see if we can find more cycles in our permutation. The letters 1, 2, 3 have already been used up, but 4 has not, so let's repeat the same game starting with 4.

   [The students work in their teams to find the path of σ starting at 4.]

4. *Beta:* 4 goes to itself; we cannot construct a cycle.

5. *Teacher:* Since we see that $4 \to 4$, we write this as (4) and call this a *trivial cycle*. It is equal to the identity function, sending every letter to itself. What do we do next?

6. *Students:* Construct the cycle starting at 5 (the next unused letter).

    [The students construct the path $5 \to 7 \to 5$ and the corresponding cycle (5 7).]

7. *Students:* Now all the numbers 1, 2, 3, 4, 5, 6, 7 have been used up—we can't construct any more cycles.

8. *Teacher:* Right. We can't and we needn't; we have now found all the cycles of our permutation. In fact, if we recall the definition of permutation product, we can see that our original permutation is actually equal to the product of the cycles we have found!

$$\begin{pmatrix} 1 & 2 & 3 & 4 & 5 & 6 & 7 \\ 6 & 1 & 2 & 4 & 7 & 3 & 5 \end{pmatrix} = (1\ \ 6\ \ 3\ \ 2)(4)(5\ \ 7)$$

    How do we know this? Take 1, for example. You can see that on both sides 1 goes to 6, and similarly for all other letters. (This is no coincidence: it's how we constructed the cycles.) Hence the permutations on the two sides are equal as functions. Notice that no number appears in two (or more) cycles on the right-hand side. The cycles are therefore said to be *disjoint*.

9. *Gamma:* Just a minute, if σ is a product of cycles, we must also take the other cycles into account when we calculate $\sigma(1)$ from this product.

10. *Delta:* Yes, but because the cycles are disjoint, 1 and 6 don't appear in any other cycle, which means that we can ignore them when calculating $\sigma(1)$.

11. *Teacher:* Right. We can summarize our work so far by saying that the permutation *has been decomposed as a product of disjoint cycles*.

    You can also check that, unlike multiplication of permutations in general, the multiplication of disjoint cycles is commutative.

12. *Epsilon:* Can we always do this? Can we decompose *any* permutation as a product of disjoint cycles?

13. *Teacher* (to the class): Well, what do you think?

14. *Epsilon:* Why shouldn't we just repeat the same process for any permutation?

15. *Alpha:* Wait a minute! What if this procedure didn't work? We were lucky that 2 went back to 1 in the first cycle, but what if it didn't? What if it went back to 6 for example? Then we wouldn't have a *cycle*.

16. *Teacher:* If we had 2 going to 6, and earlier we also had 1 going to 6, then we would have both $\sigma(1) = 6$ and $\sigma(2) = 6$. Is this possible?

17. *Beta:* No, this is impossible because a permutation is a one-to-one function so we can't have $\sigma(1) = \sigma(2)$.

18. *Teacher:* That's correct, therefore our procedure will always yield *cycles*. For the same reason, we can't have the same letter appearing in two different cycles because this too would violate the one-to-one property of the permutation [6]. This guarantees that our procedure will generate *disjoint* cycles.

19. *Teacher* (summarizes): We have jointly constructed a generic proof of the theorem: *Every permutation can be decomposed as a product of disjoint cycles.*

20. *Teacher* (moving on): As an optional homework exercise, you can try to formalize and generalize this generic proof to show that the conclusion holds for any permutation [7].

   [The teacher and students proceed to establish the uniqueness of the decomposition, but we shall skip this part because of space limitations.]

   Now we can state our full theorem: *Every permutation has a unique decomposition as a product of disjoint cycles.*

Note again the choice of generic example for the cycle decomposition theorem: a permutation on 7 letters, having cycles of lengths 1, 2, and 4. A shorter permutation on 6 letters would have been possible, with cycle lengths of 1, 2, and 3, but this orderly sequence looked to us a bit too special and possibly misleading. Thus, again, we have chosen the simplest example that would still be complex enough to represent the general case.

   This mathematical case study also demonstrates a subtle pitfall that lurks behind generic proofs. In the example, some phenomena *just happen*, automatically, but would require a proof in the general case. Thus, the cycles *just happened* to close back on the initial letter, and they also *just happened* to be disjoint. The fact that this phenomenon *just happened* in the example might conceal the need for proof in the general case, thus bypassing some of the important ideas of the general proof. In fact, this is the only place where we are using the crucial property of permutations as a *one-to-one* function. In our idealized Lakatosian dialogue, the bright students have brought up this issue themselves, but under more realistic conditions, it is more likely that the teacher would have to raise this point.

Mathematical Case Study 3: Lagrange's Theorem

Before we move on with a generic proof, we bring a brief mathematical introduction of Lagrange's theorem. We do this not by presenting a crash course in group theory but by limiting our explanations to the context of the examples used. In effect, we are preceding the generic proof of Lagrange's theorem by a *generic introduction* to the elements of group theory used in the proof.

The entire discussion of the generic proof occurs within the group Z_{12}, consisting of the set $\{0, 1, 2, \ldots, 11\}$ and the operation of *addition modulo* 12, denoted by $+_{12}$. For example, $2 +_{12} 3 = 5$, $5 +_{12} 7 = 0$, $5 +_{12} 8 = 1$, and, in general, $a +_{12} b$ is defined to be the remainder of the usual sum $a + b$ on division by 12. Z_{12} is a *group* in the sense that it contains 0 and is *closed* under addition modulo 12, i.e., if a and b are in Z_{12}, then so is $a +_{12} b$ [8]. Furthermore, Z_{12} is a *finite* group since it contains a finite number of elements.

A *subgroup* of Z_{12} is a subset of $\{0, 1, 2, \ldots, 11\}$, which is in itself a group under the operation defined in Z_{12}. For example, it can be checked that the subset $H = \{0, 3, 6, 9\}$ is a subgroup of Z_{12}, since it contains 0 and is closed under $+_{12}$. (For example, $6 +_{12} 9 = 3$, which is again a member of H.)

The *order* of a finite group G is the number of its elements, and is denoted $o(G)$. Thus, $o(Z_{12}) = 12$ and $o(H) = 4$. We note that 4 divides 12. As it turns out, this is not a coincidence, and is in fact an example of the following theorem, which is probably the most important theorem in elementary group theory.

Lagrange's theorem: *If* H *is a subgroup of a finite group* G, *then the order* H *divides the order of* G.

A generic proof of Lagrange's theorem: We will carry out the proof of the theorem for the group $G = Z_{12}$ and the subgroup $H = \{0, 3, 6, 9\}$. It may seem odd to prove that $o(H)$ divides $o(G)$ for this case, for obviously no proof is necessary for the fact that 4 divides 12. However, while we already know that $o(H)$ divides $o(G)$ in our example, we do not know *why* this is so, and to this end we do need the generic proof. The generic proof will demonstrate the general *process* that serves to carry out the proof in general.

The main idea of the proof is that by creating "shifts" of H, we obtain a partition of G into disjoint subsets that have the same cardinality as H, from which we can calculate $o(G)$ from $o(H)$. Specifically, given an element g in G, we define *the coset of* g *and* H *in* G as the following subset of G:

$$H +_{12} g = \{h +_{12} g : h \text{ in } H\}$$

We calculate the cosets of H in G in our example.

$$H +_{12} 0 = \{0 +_{12} 0, 3 +_{12} 0, 6 +_{12} 0, 9 +_{12} 0\} = \{0, 3, 6, 9\},$$
$$\text{(i.e., } H \text{ itself).}$$

Similarly:

$$H +_{12} 1 = \{1, 4, 7, 10\}$$
$$H +_{12} 2 = \{2, 5, 8, 11\}$$
$$H +_{12} 3 = \ldots$$

We leave it for the reader (or for the class in the Lakatosian scenario) to show that from now on we are not getting new cosets but are only repeating the old ones.

The number of distinct cosets of any subgroup H of a finite group G is called the *index of* H *in* G, and is denoted $i_G(H)$. In our example, where $G = Z_{12}$, we have $i_G(H) = 3$.

We can see that all the cosets have the same number of elements as H (4 in our example), and that each element in G appears in one and only one of the cosets of H in G. Since G is now presented as the disjoint union of the different cosets of H in G, we can conclude that the order of G is the number of distinct cosets of H in G times the order of H, namely:

$$o(Z_{12}) = o(H) \times i_G(H)$$

which is an even stronger statement than what we had to prove.

As in the permutation decomposition theorem, the virtual class would have noticed that some of the relations that in the example "just happened" would require a proof that they should *always happen* in the general case. This includes the facts that all cosets have the same number of elements (the order of H) and that distinct cosets are disjoint. In fact, as in the permutations example, proving these "lemmas" is where we actually use the group and subgroup definitions.

The main contribution of this mathematical case study to the general discussion of generic proving is the slippery nature of the question, "What is a good generic example in the context of a generic proof?" Indeed, the group Z_{12}, being a *cyclic* group (generated by a single element), is the simplest kind of group imaginable, and thus definitely *not* a generic example of a finite group. If someone asked you for an example of a "typical" finite group, you would definitely not think of giving them this example. Still it does a fair job of exemplifying the main ideas of the proof of Lagrange's theorem. Moreover, choosing a more "generic" example of a group (say, the so-called *symmetric group* S_3 with 6 elements or S_4 with 24 elements), would have made the generic proof computationally much more complicated, and what we might have gained in generality would have been lost in simplicity and learnability. Thus, a delicate balance between generality and simplicity is needed in making this didactical choice. When the mathematical objects making up the proof are simple (numbers, permutations, Eulerian circuits in graphs), then the balance leans toward the generality of the generic example. When the objects become more complicated and abstract (groups, subgroups, cosets), the balance leans toward simplicity rather than generality.

Finally, this mathematical case study also highlights the fact that the test of genericity should be applied not to the example itself (Z_{12} is not a good generic example of a group) but rather to the proving process that this example generates: the process of partitioning G by its cosets, and the properties of this partition, are quite general, though the group to which we are applying this process is not.

Reflections on Scope and Method

Reflecting and generalizing from the above examples brings up some important mathematical and educational issues regarding generic proofs. We list these issues as questions with tentative partial answers.

What Are the Strengths of Generic Proofs?

In learning proofs, one can distinguish two different types of activities: understanding a proof (presented by a book or teacher) and creating a proof (given the theorem). Generic examples can substantially help teachers and students in the pursuit of both goals.

First, generic proofs can help understanding by enabling students to engage with the main ideas of the complete proof in an intuitive and familiar context, temporarily suspending the formidable issues of full generality, formalism, and symbolism. Although a complete formal proof may be beyond the reach of almost all schoolchildren (Healy and Hoyles 2000; Stylianides 2007), we could imagine a classroom activity whereby even elementary schoolchildren learn the generic proof of the perfect square theorem (our mathematical case study 1) and produce their own versions for other examples. Indeed, they would most likely feel that they were carrying out *the same* proof.

In more complicated proofs, such as our permutations example, it is possible to build up the complexity gradually, via a chain of successively more elaborate *partial* generic proofs, each highlighting finer points of the proof that were not salient in previous steps. Thus, though the complete formal proof may be beyond reach for most high school students, they can still get a good view of the main ideas via a generic proof. Even for college-level students, preceding the complete proof by a generic version may help in highlighting the main ideas of the proof, separating them from the technicalities of formalism and notation.

Through the process of generic proving, teachers can help students move beyond empirical proof schemes (Harel and Sowder 1998, 2007) and raise their need for proof (Zaslavsky et al. 2012). Finally, generic proving can help students create the complete proof by serving as a graded sequence of hints in a guided discovery process. This aspect is discussed more fully later.

What Are the Weaknesses of Generic Proofs?

The main weakness of a generic proof is, obviously, that it does not really prove the theorem. The "fussiness" of the full, formal, deductive proof is necessary to ensure that the theorem's conclusion infallibly follows from its premises. In fact, some of the more subtle points of a proof are prone to be glossed over in the context of the generic proof: some steps that "just happen" in the example may require a special argument in the complete proof to explain *why* they happen and to ensure that they *always* happen. In the generic proof of the cycle decomposition theorem, for example, we have seen that cycles just "turn out" to close back to their first element and that cycles just "turn out" to be disjoint. In fact, if we had not been careful, we could have completed the generic proof without ever utilizing the crucial one-to-one property of permutations. Since these essential issues do not naturally come up in the course of generic proving, the teacher's initiative here is crucial. Similarly, in the generic proof of Lagrange's theorem, students who see that all cosets have the same number of elements may not notice that this fact requires proof and, in fact, depends crucially on the group definition.

We can capture this important difference by saying that in the generic proof some facts are simply *observed*, whereas in the complete proof they have to be *derived*. Thus, the fact that all cosets have the same number of elements (or even, for that matter, Lagrange's theorem itself) is simply observed in the case of our example but must be derived in the general proof. This issue presents a challenge to anyone who teaches with generic proofs: the teacher needs to motivate the students to learn these additional parts of the proof, which may appear unnecessary in the context of the example [9].

When We Have Presented a Class with a Generic Proof, What Have Students Learned? What Have We Proved?

First, as we have pointed out before, students have learned how to carry out the proof on *any* example, not just the one we demonstrated. They have also learned (at the level of the example) the *main ideas* of the proof. But what have we actually proved? What is the mathematical status of a generic proof? We know that a proof carried out on an example does not count as proof, but does this mean that we have actually proved *nothing*? Well, it must be admitted that at the formal level we have indeed proved nothing: no mathematical journal would accept for publication a generic proof of a new theorem (though even mathematical journals and textbooks occasionally indulge in isolated generic proofs for some propositions; see the example two paragraphs below). Still, the feeling persists that we did go a long way toward presenting the complete proof. How can we capture more explicitly the source of this feeling?

We could start with an observation: given a good generic proof, any professional mathematician (say, a specialist in the relevant topic area) could easily generalize and formalize it into a full formal proof. If asked about the difficulty of the task, she would likely describe this as a "technical exercise" (or even "trivial exercise"): it could require quite a bit of technical work, but hardly any additional insight, discovery, or creativity. (The fact that this "technical exercise" might be beyond the mathematical powers of most undergraduate mathematics majors is not relevant to the present theoretical discussion.)

We bring one example to support this observation. In his undergraduate textbook, *Abstract Algebra,* Herstein (1986) introduces the cycle decomposition theorem via a generic example (paralleling lines 1–11 in our virtual classroom scenario), concluding with the following remark:

> There is nothing special about the permutation [in our example] that made the argument we gave go through. The same argument would hold for *any* permutation [. . .]. We leave the formal writing down of the proof to the reader. (Herstein, 1986, pp. 132–33)

Remarkably, writing down the complete proof appears as Exercise 4 on p. 136, under the subheading "easier problems" (the other categories are "middle-level" and "harder" problems). This reference does not establish that writing down the complete proof (given a generic proof) is an easy exercise for most undergraduate students, in our experience it is not; it does support our claim that a professional mathematician would view this as just a technical exercise.

One way to clarify the status of a generic proof is to reconceptualize it as *a recipe for the learner on how to construct the complete proof* (a kind of closely guided discovery learning). We could once again imagine a classroom thought experiment, presenting the following stepwise teaching activity. The teacher (T) asks a student (S) to prove the cycle-decomposition theorem. S tries for a while and asks for help.

T: Okay, I'll give you a hint. T shows S how the proof is carried out on an example, something like lines 1–11 in our classroom scenario, and once again asks S to try to prove the theorem. If she still can't complete the proof, T gives her a few more hints (like the subsequent steps in our classroom scenario), until she says: now I have all the ingredients for constructing the full proof. (Let us assume for simplicity that she does have the technical expertise to deal with the formalization itself.)

On the face of it, this generic proving activity seems good only for didactical purposes and is not related to real proofs as conceived by working mathematicians, but actually there may be a stronger connection than first meets the eye. The reason is that even the working mathematician's proofs, as they appear in research journals, are far from being full formal proofs, and you might well ask (just as for generic proofs) what they have actually proved. The answer is not simple, but we could approximate it by saying that the working mathematician's proof is still a recipe for how to write the "ideal" complete proof. The unofficial implicit rules of mathematical discourse require only that you explicate the details of the proof to the extent that it can convince another expert in your field that given enough time and motivation, your sketch could be fleshed out into a full "ideal" proof.

Several mathematicians have expressed closely related views on the nature of a "working mathematician's proof." Here are two examples:

> Proving a claim is, for a mathematician, an act of producing, for an audience of peer experts, an argument to convince them that a proof of the claim exists.. . . The convinced listener feels empowered by the argument, given sufficient time, resources, and incentive, to actually construct a formal proof. (Hyman Bass, personal communication [10])

> To be sure, in practice no one actually bothers to write out such formal proofs. In practice, a proof is a sketch, in sufficient detail to make possible a routine translation of this sketch into formal proof. (Mac Lane, 1986, p. 377)

In view of these quotations, since both the professional mathematician and the students do not write a complete formal proof but only a recipe that convinces someone else that writing such a proof would be "merely a technical exercise" (or, in Mac Lane's words, "a routine translation"), it is possible to view the difference between a generic proof and a mathematician's proof as a matter of degree rather than kind.

Not All Proofs Are Equally Amenable to a Genuine Generic Version. Can We Characterize the Proofs (or Parts Thereof) That Are So Amenable?

This fascinating and difficult question raises a host of mathematical and educational issues. An answer would likely involve the form and structure of the proof (a mathematical aspect), but the effectiveness of a generic version is expressed in terms of its ability to render the main ideas of the general proof accessible to a particular audience (an educational aspect).

One observation that stands out of the mathematical case studies is that if a proof involves an act of construction (of a mathematical object or process), then this construction can be effectively presented via a generic example. Thus, we have shown how to construct all factorizations of the given number in our first mathematical case study, a cycle decomposition in our second mathematical case study, and a partition of the group into its cosets in the third. Significantly, whether a proof does or does not involve an act of construction may depend on how we choose to formulate the proof. Often an act of construction in a proof is hidden by the linear mathematical formalism but may be revealed by "structuring" the proof (Leron 1983, 1985a), whence a generic version of the proof may become accessible.

Some proofs may not seem on the surface to be amenable to a generic version because of their structure or logical form, or the nature of the mathematical objects involved, for example, proof by contradiction or proofs involving infinite objects. But even in such cases, we can often isolate some constructive element that can be presented via a generic example. We mention three such examples.

1. *Euclid's proof of the infinitude of prime numbers.* The basic construction here (given any finite set of primes, construct a new prime not in the set) can be presented via a generic example. In fact, Euclid himself does this in his *Elements,* Book IX, Proposition 20 (Reid & Knipping, 2010, p. 135; see also Leron, 1985b). Reid and Knipping point out that Euclid had no choice but to base his proof on a generic example, since he lacked the notation to discuss *any* number.

2. *Cantor's proof that the real numbers are uncountable, where given any list of real numbers, a new real number is constructed by the diagonal method.* The diagonal method itself (given a rectangular table of numbers, construct a row different from all the rows in the table) can be first introduced via small finite generic examples and then gradually extended to the infinite case (e.g., Leron and Moran 1983).

3. *Lagrange's theorem on finite groups: the order of a subgroup divides the order of the group.* Since this theorem concerns a relation on the collection of all finite groups, it might be hard to see at first glance how it could be helped by a generic example. But since the *proof* involves a construction (a partition of the group into cosets), it is possible to devise classroom activities that demonstrate the main ideas of the proof on a generic example, as we show in our third mathematical case study.

In contrast to these examples, we may consider the Heine-Borel theorem from analysis:

4. *A subset of R^n is compact if and only if it is closed and bounded.* The mathematical concept of compact (or closed or bounded) space can of course be exemplified, but since the theorem and its proof deal with complex logical relations between infinite collections of infinite objects, it is hard to imagine how it could be effectively demonstrated through an example.

Conclusion

At the core of this article is the method of generic proving: accessing a complicated proof by a chain of intermediate steps, where each step highlights some of the ideas of the proof by performing them on a generic example, and where details, refinements, and complications are gradually added as we progress along the chain. Ideally, this process should enable the learner to reach even a complicated and "unnatural" proof via a sequence of relatively easy and natural steps.

We conclude with a few possible directions for further theoretical and empirical study.

- Elaborate on methods of using generic proofs in actual classrooms, both for understanding and for generating proofs, and test their efficacy empirically.

- Study the work of college students to investigate the hypothesis that preceding a complex formal proof by a generic version would enhance their understanding of the main ideas behind the proof.
- Interview professional mathematicians to investigate the role of generic proofs in their research and teaching.
- Interview professional mathematicians to investigate the hypothesis that proofs are stored in their long-term memory in the form of generic examples. On theoretical grounds, two supporting arguments can be given for this hypothesis. First, this is simply a more economical way to store such proofs in memory: one only memorizes the main ideas, leaving out the technical details, which can be readily reconstructed (by an expert) when needed. Second, this is an efficient way of storing past experiences so that one may recognize similar problem-solving situations in the future and reuse the same methods [11].

Notes

[1] A comprehensive discussion of research on proofs in mathematics education, including generic proofs, is beyond the scope of this article. The reader is referred to the excellent book by Reid and Knipping (2010) for such a survey.

[2] Readers are referred, for example, to Balacheff's (1988) distinctions between pragmatic vs. conceptual proofs, and between generic example vs. thought experiment and to Herbst's (2004) discussion of students' interaction with drawings and diagrams in geometrical proofs, and the teacher's role in supporting this interaction.

[3] In adopting this format, we have obviously been influenced by the imaginary classroom scenario in Lakatos's seminal *Proofs and Refutations* (1976), as well as the dialog with "the ideal mathematician" in Davis and Hersh (1981). As in Lakatos (1976), the scenario is taking place in an *idealized class* of bright and highly motivated students. A realistic class is a noisy situation (pun intended), and we prefer to describe our ideas first in a simplified and idealized setting. A realistic scenario would be much lengthier and meandering, with lots of false starts and dead ends and, most significantly, with the teacher having to shoulder a larger portion of the classroom discussion, rather than eliciting most of the insights from the students.

[4] The numbers on which the permutation operates could be any n symbols and are therefore referred to as letters.

[5] In this graphical representation of the permutation, the numbers in the bottom row are the images of the corresponding numbers in the top row. Thus, for the permutation σ defined next, $\sigma(1) = 6$, $\sigma(2) = 1$, $\sigma(3) = 2$, etc.

[6] A little trick is needed here, in fact, a masked application of mathematical induction: assume on the contrary that there is a letter that appears in two different cycles, and take *the first* such occurrence. Then the letters appearing immediately before these two occurrences must be different, which cannot happen in a one-to-one function.

[7]   For one such proof see Gallian (1990, p. 88).

[8]   The general definition of a group includes more requirements, namely associativity and the existence of inverses. However, we do not need to worry about these here because, in general, associativity for addition mod *n* can be shown to be "inherited" from the associativity of the usual addition of integers, and the existence of inverses can be shown, in the finite case, to follow from the other group properties.

[9]   This is similar to the teacher's problem when proving a theorem in geometry, which appears obvious from the accompanying figure or from dynamic geometry activities.

[10] From a draft manuscript by Bass for a book in preparation under the editorship of Peter Casazza, Steven G. Krarb, and Randi D. Ruden, currently entitled "I, Mathematician."

[11] This last argument sits well with Minsky's K-Theory (1985, Chapter 8), according to which the most efficient way for people to store their memories for future problem solving is at middle-level abstraction: not too concrete but also not too abstract. It also sits well with Rosch et al.'s (1976) theory of basic level categories: "[Categories] within taxonomies of concrete objects are structured such that there is generally one level of abstraction at which the most basic category cuts can be made. In general, the basic level of abstraction in a taxonomy is the level at which categories carry the most information, possess the highest cue validity, and are, thus, the most differentiated from one another" (p. 383).

References

Balacheff, N. (1988). "Aspects of proof in pupils' practice of school mathematics." In Pimm, D. (Ed.) *Mathematics, Teachers and Children,* pp. 216–35. London: Hodder and Stoughton.

Davis, P. J., and Hersh, R. (1981). *The Mathematical Experience,* Boston: Houghton Mifflin.

Gallian, J. A. (1990). *Contemporary Abstract Algebra* (2nd edition), Lexington, MA: D. C. Heath & Co.

Harel, G., and Sowder, L. (1998). "Students' proof schemes: Results from exploratory studies." In Schoenfeld, A. H., Kaput, J., and Dubinsky, E. (Eds.) *Research in Collegiate Mathematics Education III,* pp. 234–83. Providence, RI: American Mathematical Society/ Mathematical Association of America.

Harel, G., and Sowder, L. (2007). "Toward comprehensive perspectives on the learning and teaching of proof." In Lester, F. (Ed.) *Second Handbook of Research on Mathematics Teaching and Learning,* pp, 805–42. Charlotte, NC: Information Age Publishing.

Healy, L., and Hoyles, C. (2000). "A study of proof conceptions in algebra." *Journal for Research in Mathematics Education* **31**(4), 396–428.

Herbst, P. (2004). "Interactions with diagrams and the making of reasoned conjectures in geometry." *ZDM* **36**(5), 129–39.

Herstein, I. N. (1986). *Abstract Algebra,* New York: Macmillan.

Lakatos, I. (1976). *Proofs and Refutations: The Logic of Mathematical Discovery.* Cambridge, UK: Cambridge University Press.

Leron, U. (1983). "Structuring mathematical proofs." *American Mathematical Monthly* **90**(3), 174–85.

Leron, U. (1985a). "Heuristic presentations: The role of structuring." *For the Learning of Mathematics* **5**(3), 7–13.

Leron, U. (1985b). "A direct approach to indirect proofs." *Educational Studies in Mathematics* **16**(3), 321–25.

Leron, U., and Moran, G. (1983). "The diagonal method." *The Mathematics Teacher* **76**(9), 674–76.

Mac Lane, S. (1986). *Mathematics: Form and Function,* New York: Springer-Verlag.

Malek, A., and Movshovitz-Hadar, N. (2011). "The effect of using transparent pseudo-proofs in linear algebra." *Research in Mathematics Education* **13**(1), 33–58.

Mason, J., and Pimm, D. (1984). "Generic examples: Seeing the general in the particular." *Educational Studies in Mathematics* **15**(3), 277–89.

Minsky, M. (1985). *The Society of Mind,* New York: Simon & Schuster.

Movshovitz-Hadar, N. (1988). "Stimulating presentations of theorems followed by responsive proofs." *For the Learning of Mathematics* **8**(2), 12–19, 30.

Reid, D. A., and Knipping, C. (2010). *Proofs in Mathematics Education: Research, Learning and Teaching,* Rotterdam, Netherlands: Sense Publishers.

Rosch, E., Mervis, C. B., Gray, W. D., Johnson, D. M., and Boyes-Braem, P. (1976). "Basic objects in natural categories." *Cognitive Psychology* **8**(3), 382–439.

Rowland, T. (1998). "Conviction, explanation and generic examples." In Olivier, A., and Newstead, K. (Eds.) *Proceedings of the 22nd Conference of the International Group for the Psychology of Mathematics Education,* Vol. 4, pp. 65–72. Stellenbosch, South Africa: University of Stellenbosch.

Stylianides, A. J. (2007). "Proof and proving in school mathematics." *Journal for Research in Mathematics Education* **38**(3), 289–321.

Zaslavsky, O., Nickerson, S. D., Stylianides, A. J., Kidron, I., and Winicki- Landman, G. (2012). "The need for proof and proving: Mathematical and pedagogical perspectives." In Hanna, G., and de Villiers, M. (Eds.) *Proof and Proving in Mathematics Education,* pp. 215–29. Dordrecht, Germany: Springer.

Extreme Proofs I: The Irrationality of $\sqrt{2}$

John H. Conway and Joseph Shipman

Mathematicians often ask, "What is the best proof" of something, and indeed Erdős used to speak of "Proofs from the Book," meaning, of course, God's book. Aigner and Ziegler (1998) have attempted to reconstruct some of this Book.

Here we take a different and more tolerant approach. We shouldn't speak of "the best" proof because different people value proofs in different ways. Indeed one person's value might oppose another's. For example, a proof that quotes well-known results from Galois theory is valued negatively by someone who knows nothing of that theory but positively by the instructor in a course on Galois theory. Other "values" that have been proposed include brevity, generality, constructiveness, visuality, nonvisuality, "surprise," elementarity, and so on. A single mathematician may hold more than one of the values dear. Clearly the ordering of proofs cannot be a total order.

It is enjoyable and instructive to find proofs that are optimal with respect to one or more such value functions not only because they tend to be beautiful but also because they are more likely to point to possible generalizations and applications.

In this respect, we can discard a proof C that uses all the ideas of shorter proofs A and B because nobody should value it more highly than both A and B. We model this by putting C on the line segment AB, and it suggests that we think of proofs of a given result as lying in a convex region in some kind of space, which in our pictures will be the Euclidean plane.

Indeed, because at any given time there are only finitely many known proofs, we may think of them as lying in a polyhedron (in our pictures, a polygon), and the value functions as linear functionals, as in optimization theory, so that any value function must be maximized at

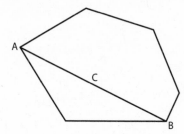

FIGURE 1. A visualization of "proof space."

some vertex. We shall call the proofs at the vertices of this polygon the *extreme proofs.*

It can be difficult to decide whether two proofs are "really the same." Usually a proof has a natural domain of applicability, which we shall call its "scope," and this domain provides us with the "scope test," a necessary condition for essential difference: proofs are "really different" if they have different scopes.

The Irrationality of $\sqrt{2}$

This article explores the proof space of one of the oldest and most familiar theorems of all: that there is no rational number a/b whose square is 2. This theorem was traditionally credited to Pythagoras, but it is perhaps more correct to ascribe it to the Pythagorean school, for Iamblichus tells us of the rule that all discoveries of this school were attributed to its founder.

According to Plato's *Theaetetus*, Theodorus demonstrated how to prove the irrationality of square roots of nonsquare numbers up to 17, an assertion that has given rise to much speculation about the reason he stopped there. It has been suggested that the reason might be that the proofs were geometrical and differed from case to case. Our first proof is one that Theodorus might have used for $\sqrt{2}$.

Tennenbaum's "Covering" Proof

The assertion that $\sqrt{2}$ is the quotient p/q of two integers is equivalent to $p^2 = 2q^2$, which is equivalent to the assertion that a $p \times p$ square has the same area as two $q \times q$ ones.

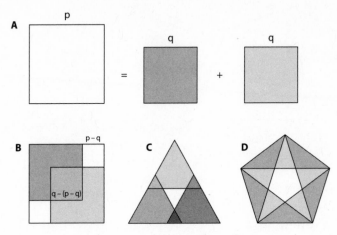

FIGURE 2. "Covering" proofs.

Stanley Tennenbaum therefore supposes that the large square at left in Figure 2A is the smallest one of integral side (p) that has the same area as two smaller ones of equal integer side (q) at right. Then, fitting the two smaller squares into opposite corners, we obtain Figure 2B, in which the central doubly covered square has the same total area as the two uncovered ones, a smaller example than we started with—a contradiction.

DISCUSSION

Traditionally, this type of argument is called a "proof by descent," and it relies on the principle (usually tacitly assumed by the ancients) that any nonempty set of positive integers has a least element.

This type of proof is difficult to generalize, but a glance at Figure 2C suggests a similar proof for $\sqrt{3}$, and you might find one for $\sqrt{5}$ after studying Figure 2D.

"Folding" Proofs

A traditional Greek statement of our theorem is that the diagonal and side of a square are incommensurable; that is, they cannot both be

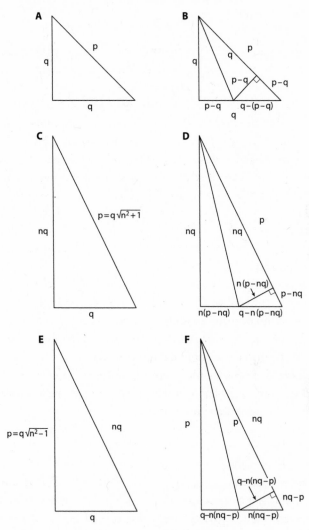

FIGURE 3. "Folding" proofs.

integer multiples of a common ("unit") length. Suppose that they are, and let the smallest such pair of integers be *p* and *q*, as in Figure 3A (in modern terms, $\sqrt{2} = p/q$). Folding the half-square triangle as in Figure 3B, we see that there is a smaller half-square triangle with integer sides—contradiction.

Discussion

In some sense, this proof is mechanically the same as Tennenbaum's; both suppose that the fraction in least terms that represents $\sqrt{2}$ is p/q and deduce the same contradiction, that $(q - (p - q))/(p - q)$ is a smaller one. It also assumes basic facts about Euclidean geometry, including the Pythagorean theorem. In effect, both proofs carry out geometrically the first step in the division algorithm that is at the heart of the Unique Factorization Theorem (also known as the Fundamental Theorem of Arithmetic).

However, our "scope test" shows that this proof is more general than the covering proof because essentially the same proof works for showing the irrationality of square roots of numbers of the form $n^2 + 1$.

As before, all labeled sides have integer lengths, and the fold shows a smaller triangle with integer-length sides that is similar to the original one. Apostol (2000) provided a similar proof.

Just as we needed a geometric equivalent to the division algorithm, to perform this proof in the style of Euclid requires the geometrical equivalent of the fact that any set of positive integers has a smallest element, namely, the Archimedean axiom that given any two segments, some multiple of each segment exceeds the other in length.

An analogous proof handles the case of $\sqrt{(n^2 - 1)}$ (as was also noted by Apostol): use a right triangle with base q, hypotenuse qn, and height $p = q\sqrt{(n^2 - 1)}$, and make the "same fold."

We thus obtain proofs for the irrationality of the square roots of 2, 3, 5, 8, 10, 15, 17, 24, 26, 35, 37, 48, 50, 63, 65, 80, 82, 99, . . . (of course, 8 already follows from 2).

We can also handle 6 because $\sqrt{6} = (1/2)\sqrt{24}$, and 7 because $\sqrt{7} = (1/3)\sqrt{63}$, and $\sqrt{11}$ because $\sqrt{11} = (1/3)\sqrt{99}$.

We observe that this trick works for all nonsquares D because the Pell equation $x^2 - 1 = Dy^2$ is always solvable in integers (Stark 1978). This is not true for the form $x^2 + 1$ in some cases, for example, when $D = 7$, so it was necessary to generalize the proof to handle $\sqrt{(n^2 - 1)}$.

No surviving manuscripts indicate that the ancient Greeks knew this fact (although we would not put it past Diophantus or Archimedes). They were certainly capable of finding the "geometric proof" that we now know always exists, in any particular case. Thus, Theodorus could have had a geometric proof in the style of Euclid for all D.

For $D = 13$, the smallest solution is $(649^2 - 1 = (13)(180)^2)$, so the original "$n^2 + 1$" proof works best $(18^2 + 1 = (13)5^2)$. For $D = 14$, we can use $(15^2 - 1 = (14)4^2)$. This answer provides a possible explanation for why Theodorus stopped at 17; to do 19 requires finding $(170^2 - 1 = (19)(39)^2)$. It would have been even more difficult to demonstrate proofs for $D = 31$ $(1520^2 - 1 = (31)(273)^2)$, and for $D = 61$, where the smallest solution is $(29{,}718^2 + 1 = (61)(3805)^2)$.

"Traditional" Even/Odd Proof

The traditional arithmetical proof is to suppose that $\sqrt{2} = p/q$ in least terms, so that $p^2 = 2q^2$. But the square of an odd number is odd, so p must be even; say that it equals $2r$. Now $4r^2 = 2q^2$, and we deduce $q^2 = 2r^2$, showing that 2 equals the "simpler" fraction q/r.

Discussion

This proof is "extreme" in seeming to depend on the most elementary concepts. We did need the fact that fractions have canonical "least" forms, implying that any set of positive integers has a smallest element. Also, we are relying on the division of integers into two classes, "even" and "odd," with the property that a product is odd iff both factors are odd:

*	even	odd
even	even	even
odd	even	odd

How much does this proof generalize? Note that the same argument works for higher-order roots because the table implies that any power of an odd number is odd, not just its square.

We can replace 2 by any prime, and the proof is the same. Call a number "3even" (pronounced "threeven") if it is divisible by 3, otherwise it is "3odd" ("throdd"), for instance; then

*	3even	3odd
3even	3even	3even
3odd	3even	3odd

Of course, in using the fact that a prime divides a product only if it divides one of the factors, we are assuming the key lemma necessary to prove the Unique Factorization Theorem; this lemma is obvious for the prime 2, but we didn't totally avoid "unique factorization."

This proof doesn't work for nonprimes (2 and 3 are 6odd, but their product isn't). To generalize to roots of nonprimes, we need to make a bigger table with more residue classes, but that's not really the "same proof" any longer.

Bashmakova's Proof

We don't know whether Theodorus stopped before or after proving the theorem for 17, and he apparently could not handle the general case (in the dialogue, Theaetetus claims to have improved on Theodorus by proving the general case). But here is a proof that works "up to 17," according to the Russian historian of mathematics, Isabella Bashmakova (Bashmakova and Lapin 1986):

Suppose that $p^2 = Nq^2$, with p/q in least terms. If N is divisible by 4, we may replace N by $N/4$ until it isn't. Suppose that N is even but not divisible by 4: then p^2 is even, so p is, so $p^2/2$ is, so $(N/2)q^2$ is, so (since $N/2$ is odd) q^2 is, so q is, so p/q is not in least terms—contradiction. So assume that N is odd. Since at least one of p and q must be odd, both are, so p^2 and q^2 are both 1 mod 8, so N is also. Thus we obtain a contradiction for any N except 1, 9, 17, 25, 33, . . . of which 17 is the first nonsquare and so the first failure of the proof.

Discussion

In one sense, this is not an "extreme" proof because it treats the cases of even and odd N with different ideas, and in fact for $N = 2$ it is really the same as the "traditional" proof. We include it for historical interest and because the "remainder" argument is a distinctly new idea, even though it only applies for N equal to 3, 5, or 7 mod 8.

Reciprocation Proof

This proof (from Conway and Guy 1996) overcomes the Unique Factorization difficulty. Suppose again that $\sqrt{2} = P/Q$. Then it also equals $2/\sqrt{2} = 2Q/P$. These two numbers P/Q and $2Q/P$ have "fractional parts" expressible as q/Q and p/P, which must be equal. But then P/Q and p/q

must be equal, so that p/q is the desired simpler fraction because $p < P$ and $q < Q$.

DISCUSSION

This solution handles $\sqrt{N}$ for any nonsquare N without invoking Unique Factorization; however, it does use "division with remainder" once; this much is unavoidable if the notion of "fractional part" is to make sense. Euclid (*Elements*) also provides a proof (Book X, Proposition 9) with this "scope."

Unique Factorization Proof

Let a and b be positive integers. There is a prime factorization of a^2 (obtained by doubling the exponents in a prime factorization of a) in which the exponent of 2 is even, and similarly for b^2; but then $2b^2$ has a factorization in which the exponent of 2 is odd, and, by the Fundamental Theorem of Arithmetic, different factorizations must be of different numbers, so a^2 cannot equal $2b^2$, and a/b is not a square root of 2.

DISCUSSION

This proof uses a hammer to crack a nut. It is not self-contained because we have not proven the Fundamental Theorem of Arithmetic. It clearly generalizes to show that no rational number can be the nth root of an integer that is not a perfect nth power: for the exponents of the primes in the factorizations of a^n and b^n are divisible by n, and if a^n is to equal kb^n, then all the exponents in the prime factorization of k must also be divisible by n, and when you divide them all by n you obtain an integer whose nth power is k. How is this proof "extreme"? Well, it's the shortest proof and the most transparent proof if the Fundamental Theorem of Arithmetic "comes for free," and it generalizes to arbitrary integers in both the base and the exponent.

After explaining how he had shown the irrationality of square roots, Theaetetus remarked "and the same for solids," suggesting that he also had a way to handle cube roots. Wayne Aitken has noted (FOM [Foundations of Mathematics] e-mail discussion list archived at http://www .cs.nyu.edu/pipermail/fom/2007-November/012259.html) that, even though Euclid never stated the Fundamental Theorem of Arithmetic, one may use his proposition VIII.8 to derive the irrationality of nth roots by an argument similar to Euclid X.9, which treats square roots.

Analytic Proof

This proof, presented in Laczkovich (2001), is a quickie for those who know some algebra. For positive integers n, $(\sqrt{2} - 1)^n$ has the form $a\sqrt{2} + b$ where a and b are integers, not necessarily positive. But if $\sqrt{2}$ were rational with denominator D, then for integral a,b, $a\sqrt{2} + b$ would be rational with denominator dividing D and so could not approach a limit of 0 as $(\sqrt{2} - 1)^n$ must, since $0 < (\sqrt{2} - 1) < 1$.

Discussion

From this type of argument, one may learn not only that $\sqrt{2}$ is irrational but also some quantitative information on how closely it can be approximated by rational numbers. The proof obviously generalizes to other

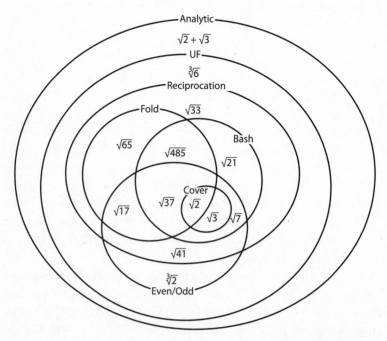

FIGURE 4. Relationships between proofs, with examples of all possibilities.

square roots; only slightly less obviously, it generalizes to "kth roots," using expressions of the form $aN^{(k-1)/k} + bN^{(k-2)/k} + \ldots + yN^{1/k} + z$, which will have denominators no larger than D^{k-1} if $N^{1/k}$ is rational with denominator D.

In fact, the proof generalizes further still and shows that all real algebraic integers (roots of polynomials with integer coefficients and a highest-degree coefficient of 1) are either integers or irrational numbers. All you have to note is that higher powers of the root can be replaced recursively by sums and differences of lower powers because the root satisfies a monic polynomial.

Because we took advantage of a new principle, that every real number is "close to an integer" (a distance of at most $1/2$), this generalization is even more "extreme" than the one from the Unique Factorization proof.

Conclusion

All these proofs show the irrationality of various numbers; in the order we have presented them, each one handled some new cases to which the previous proofs did not apply. We summarize the seven proofs in the table that follows and provide a diagram illustrating their "scopes."

TABLE 1. Synopsis of Proof Types

Name of Proof	Domain of Applicability	Key Idea	Remarks	Why Proof Is "Extreme"
Covering	$\sqrt{2}, \sqrt{3}, \sqrt{?}$	Doubly covered = uncovered	$n = 2$ case from Stanley Tennenbaum; $n = 3$ from Conway	Visually obvious
Folding	Square roots of $(n^2 + 1)$	Folding a triangle	Apostol (2000): Similar proof for $(n^2 - 1)$. Conway and Shipman: handles all integers by Pell equation theory	Purely geometrical

TABLE 1. (*Continued*)

Name of Proof	Domain of Applicability	Key Idea	Remarks	Why Proof Is "Extreme"
Traditional even/odd	All roots of primes	Even–odd argument	Who first noted that this works for any prime?	Depends on simplest concepts
Bashmakova's	Square roots of integers ≠ 1 mod 8	Remainder argument	Isabella Bashmakova and A. I. Lapin (1986) noted connection to *Theaetetus'* Theodorus (Plato)	Historical interest?
Reciprocation	Square roots of all integers	1 Step of division algorithm	Conway and Guy (1996)	Purely arithmetical; "slickest"
Unique factorization	All roots of all integers	Compare exponents in factorizations	Who first stated that all roots of nonpowers are irrational?	Most useful, shortest if Unique Factorization Theorem assumed
Analytic	Algebraic integers	Analytic estimate	Laczkovich [L] gives proof for $\sqrt{2}$, generalizes to roots of integers	Most general, quantitative, "surprising"

References

Apostol, T., "Irrationality of the Square Root of Two—A Geometric Proof," *American Mathematical Monthly* 107: 841–42, 2000.

Aigner, Martin, and Günter M. Ziegler, *Proofs from THE BOOK*, Springer, Berlin, 1998.

Bashmakova, I. G., and A. I. Lapin, *Pifagor*, Kvant, No. 1, 1986, p. 10 (in Russian).

Hardy, G. H., and E. M. Wright, *An Introduction to the Theory of Numbers*, 4th ed., Clarendon Press, Oxford, U.K., 1959.

Conway, John H., and Richard K. Guy, *The Book of Numbers*, Copernicus, New York, 1996.

Euclid, *Elements: All Thirteen Books in One Volume*, T. L. Heath (trans.), Green Lion Press, Santa Fe, NM, 2002, http://www.greenlion.com/euclid.html.

Laczkovich, Miklós, *Conjecture and Proof*, Mathematical Association of America, Washington, DC, 2001.

Plato, *Theaetetus*. Benjamin Jowett (trans.), 3rd ed., Vol. 4 of 5, Oxford University Press, 1892, Oxford, U.K., http://oll.libertyfund.org/title/768/93834.

Stark, Harold M., *An Introduction to Number Theory*, MIT Press, Cambridge, MA, 1978.

Stuck in the Middle:
Cauchy's Intermediate Value Theorem and the History of Analytic Rigor

MICHAEL J. BARANY

Intermediate Values

With the restoration of King Louis XVIII of France in 1814, one revolution had come to an end, but another was just beginning. Historians often describe the French Revolution of 1789, along with its reactions and repercussions, as the start of the modern era. For many historians of mathematics, however, the modern era began with the Bourbon Restoration and the mathematics of Augustin-Louis Cauchy.

In a 1972 talk at a Mathematical Association of America sectional meeting, Judith Grabiner offered an important interpretation of this critical juncture in the history of mathematics [5]. Asking whether mathematical truth was time-dependent, Grabiner argued that while truths themselves may not depend on time, our knowledge certainly does. She stressed the revolutionary character of Cauchy and his contemporaries' efforts to set mathematical analysis on a rigorous footing, which required applying a fundamentally new point of view to the problems and methods of their eighteenth-century forebears. One of Grabiner's leading illustrations of this point, both in her talk and in her 1981 book on Cauchy's calculus [6], is Cauchy's proof of the intermediate value theorem, that a continuous real-valued function f on an interval $[a, b]$ assumes every value between $f(a)$ and $f(b)$ on that interval.

Today the intermediate value theorem (IVT) is one of the first theorems about functions that advanced undergraduates learn in courses on mathematical analysis. These courses in turn are often the first places such students are comprehensively taught the methods of rigorous proof at the heart of contemporary mathematics. Here the theorem and its proof exemplify several important aspects of rigorous analysis. At first glance, the theorem seems obvious. Indeed, generations of mathematicians before Cauchy thought its idea so obvious as not to need explicit statement or justification. On the other hand, that the theorem can be proved with just some simple notions about continuity, convergence, and the system of real numbers is something quite remarkable. Students learn to take nothing for granted, to proceed systematically, and to use approximations and limiting principles to turn vague intuitions about the nature of functions into indubitable theorems.

Cauchy's undergraduate course in analysis at the École Royale Polytechnique, which he began teaching in 1816, was among the first to include a proof of the intermediate value theorem. His 1821 textbook [4] (recently released in full English translation [3]) was widely read and admired by a generation of mathematicians looking to build a new mathematics for a new era, and his proof of the intermediate value theorem in that textbook bears a striking resemblance to proofs of the theorem that continue to be taught almost two centuries later.

With this in mind, you might be surprised to learn that the theorem was proved *twice* in Cauchy's textbook, and that his more famous proof was relegated to an appendix, while Cauchy's main proof has been mostly forgotten by mathematicians and historians alike. Putting these two proofs side by side, we can add a new dimension to Grabiner's story by asking just what rigorous analysis meant for Cauchy at the dawn of modern mathematics. When Cauchy's language and methods are carefully dissected, he begins to look less like a far-sighted revolutionary who simply saw profound new meanings in old results. Instead, I argue (see also [1]) that Cauchy was "stuck in the middle": struggling to reclaim what he saw to be a neglected approach to mathematics while (perhaps inadvertently) pushing mathematicians toward a particular understanding of analytic rigor that would help define their future.

A Mathematical Revolution

Underneath the slogans of liberty, equality, and fraternity and behind the barricades and the bluster of the French Revolution, there was a massive transformation in the organization of the French state and society. For the world of mathematics, these transformations meant that, for the first time, a large cadre of elite military and civil engineers began to receive a common training in Paris in the most advanced mathematics of the day. These engineers took their mathematics and applied it to the pressing problems of the modern world: mass infrastructure, navigation, mining, energy, and war. The flagship institution where these students learned to draw, compute Taylor expansions, and see the world through mathematical eyes was the École Polytechnique (renamed the École Royale Polytechnique after Napoleon's defeat and the monarchy's return), and it was there that Cauchy made his mark as a student and then as an instructor.

Despite his acclaim beyond the walls of the École, Cauchy was not the most popular instructor among either his students or his fellow faculty. He regularly overran his allotted lecture time; his courses could be dense and difficult to follow; he revised the curriculum with abandon, disregarding the pleas of those teaching courses for which his was a prerequisite. His foes among the faculty grumbled that Cauchy, a devout Catholic and staunch supporter of the monarchy, was a bitter reactionary who owed his job more to the changing winds of national politics than to his brilliance as a teacher.[1]

In Cauchy's view, however, a restoration was just as much due for mathematics as it had been after the regrettable revolution in France, and it was no use arguing with the misguided mathematical Jacobins or Bonapartists who would have it otherwise. When Cauchy looked at the mathematics of the eighteenth century, he saw a discipline that had lost its discipline. Undoubtedly, the century had witnessed a host of marvelous mathematical innovations, but at what cost? Mathematicians such as Leonhard Euler freely toyed with nonconvergent series and ungrounded formal expressions and did not bat an eyelash when these produced absurd conclusions. Amid the swirl of infinities, heuristics, imaginary numbers, and more, it was hard to know what to believe.

At stake for Cauchy was the proper relationship between algebra and geometry. Geometry, as most saw it, was the ancient and noble science of magnitudes initiated by the Greeks and particularly associated with Euclid.[2] On the one hand, geometry referred to a specific body of problems and techniques associated with shapes and magnitudes. On the other hand, however, geometry was an emblem of philosophical exactitude and precision. Mathematicians and philosophers alike sought to proceed in the *more geometrico*, or geometric way, stating their assumptions carefully and reasoning systematically in order to produce results with absolute certainty.

Algebra, like geometry, could refer to a body of problems and techniques. From Viète and Descartes to Laplace and Lagrange, mathematicians (not all of them French) had developed the symbolic methods of algebra into a powerful tool for studying a wide range of mathematical phenomena; including those traditionally associated with geometry.[3] Also, like geometry, algebra had an implicit philosophical meaning that tied it to the unrestrained pursuit of mathematical ideas, regardless of whether each individual step had a clear geometric or physical interpretation. Eighteenth-century mathematicians saw in algebra a versatile tool for obtaining deep understandings of the world around them.

Algebra and geometry thus represented competing values. Algebraic mathematicians valued the profound mathematical truths their methods could reveal and ridiculed geometric mathematicians for their overzealous commitment to tedious proofs at the expense of vital creativity. Geometric mathematicians like Cauchy, by contrast, reviled the monstrosities that algebraic mathematics occasionally produced and sought protection in the rigorous certainties of their methods. Cauchy's textbook famously declared his desire to give his methods "all the rigor one requires in geometry, in such a way as never to resort to reasons drawn from the generality of algebra." Rarely, of course, were the values of algebra and geometry so sharply delineated. It was often in polemical writings, such as Montucla's monumental history of mathematics [7, e.g., pp. 11, 270] or the introduction to Cauchy's textbook, rather than in everyday mathematical work or in the École's courses on drawing and practical mathematics, that these stakes loomed large. Nevertheless, the tension was real, and (at least as Cauchy saw it) algebra was winning.

Two Proofs

Cauchy's textbook introduces the intermediate value theorem by noting a "remarkable property of continuous functions of a single variable": that they can represent the geometric ordinates of continuous curves [4, p. 43]. Contrary to our present emphasis on the theorem in terms of the analytic properties of continuous functions, for Cauchy the theorem is foremost about the relationship between functions and

geometry. This, we shall see, was not just his motivation but the central
idea in his proof.

The theorem's statement is recognizable to readers today, even if the
precise wording and notation appear unusual:

Theorem (Cauchy's IVT): If the function $f(x)$ is continuous with
respect to the variable x between the limits $x = x_0$ and $x = X$ and
if b designates a quantity between $f(x_0)$ and $f(X)$, one can always
satisfy the equation

$$f(x) = b$$

for one or several real values of x between x_0 and X.

But Cauchy's main proof of the theorem looks nothing like the proof
we now associate with him. Here is a rather literal translation:

Proof. To establish the preceding proposition, it suffices to see that
the curve whose equation is

$$y = f(x)$$

meets one or more times the straight line whose equation is

$$y = b$$

in the interval between the ordinates that correspond to the ab-
scissas x_0 and X. Yet it is clear that this will take place under our
hypotheses. Indeed, as the function $f(x)$ is continuous between the
limits $x = x_0$ [and] $x = X$, the curve whose equation is $y = f(x)$ passing first through the point with coordinates $x_0, f(x_0)$ and second
through the point with coordinates X and $f(X)$ will be continuous
between these two points; and, as the constant ordinate b of the
line whose equation is $y = b$ is found between the ordinates $f(x_0)$
and $f(X)$ of the two points under consideration, the line [corre-
sponding to $y = b$] will necessarily pass between these two points
so that it cannot avoid crossing the above-mentioned curve [cor-
responding to $y = f(x)$] in the interval.

The first thing to notice is that, although Cauchy employs several
variables and equations, he uses these symbolic expressions purely to
describe curves and lines in a plane. There are no algebraic manipulations whatsoever, much less sequences, bounds, or limits. He presents

the continuous function $f(x)$ as an unbroken curve connecting two points, and his proof hinges on a claim that a level line corresponding to the desired intermediate value must cross this curve. The argument is vague and unsystematic by our standards. Even though Cauchy has just given a definition of continuity,[4] his proof makes no use of it. Instead, the notion of continuity in this proof means simply that the function's corresponding curve remains unbroken.

Was Cauchy sloppy, lazy, or inconsistent with this proof? I have found nothing to suggest that he or his contemporaries had second thoughts about it.[5] Instead, we should see this as evidence of Cauchy's faith in geometric reasoning and his lingering distrust of algebra. Arguments based on unbroken planar curves were sensible and trustworthy to Cauchy in a way that arguments based on symbols and equations were not. Because he could visualize two curves crossing, he needed no further argument to establish his theorem. Here the "rigor of geometry" involved not just careful systematic reasoning but also the use of a fundamentally geometric argument.

Cauchy's more famous proof of the intermediate value theorem comes in an appendix on solving equations numerically. Here the above theorem becomes a corollary to the first theorem of the appendix, which states that if a function is continuous between $x = x_0$ and $x = X$ and if $f(x_0)$ and $f(X)$ have opposite signs, then there is at least one root satisfying $f(x) = 0$ between x_0 and X. One applies this result to the function $f(x) - b$ to obtain the familiar theorem.

Grabiner [5, p. 362] is among many who note that the proof in Cauchy's appendix is based on a method of approximating roots that was well known in Cauchy's time. Cauchy supposes that the interval between x_0 and X has length h and divides the interval into m parts of length h/m for some m greater than 1. Inspecting values of f for consecutive terms of this sequence and picking one pair of such terms where the corresponding values of f have opposite signs, Cauchy then subdivides this new interval of length h/m into m parts of length h/m^2 and repeats the process to produce two sequences of x values. The first, denoted $x_0, x_1, x_2, \ldots$, is increasing, and the second $(X, X', X'', \ldots)$ is decreasing, with corresponding terms in the two sequences coming closer and closer together.

From this process, Cauchy concludes that the sequences have a common limit a. Without citing his earlier theorem that continuous functions map convergent sequences to convergent sequences, he then simply stipulates that the sequences

$$f(x_0), f(x_1), f(x_2), \ldots$$

and

$$f(X), f(X'), f(X''), \ldots$$

must both converge to $f(a)$. Finally, Cauchy claims that, because corresponding terms of these two sequences have opposite signs, $f(a)$ must have the value 0.

At first glance, this is a pure example of the rigorous algebraic analysis for which Cauchy is known today. Nevertheless, we can still see Cauchy's preference for geometric reasoning. On the one hand, Cauchy's proof does more than one would expect if the goal were simply to prove the existence of a root. Why, for instance, carry out the argument with an arbitrary value of m when simply halving the interval each time is sufficient? Cauchy's rhetoric makes it clear that he continues to see his procedure primarily as a tool of approximation rather than as an existence proof. Thus he makes repeated reference to the possibility of there being multiple roots and elaborates on this point in two of the three scholia that follow the proof. The first such scholium is even more directly about approximation: it notes that the average of the terms x_n and $X^{(n)}$ is at most a distance $\frac{1}{2}(X - x_0)/m^n$ from the desired root a—an observation that is extraneous (in the context of Cauchy's argument) to the question of whether such a root exists but is important when one cares about rates of convergence for approximations.

Why, for that matter, did Cauchy try to find a root instead of an arbitrary intermediate value b? The proof would need few modifications to fit this more general case. Making this theorem about roots rather than arbitrary values allowed Cauchy to preserve it as an argument about a curve meeting a line (in this case, the x-axis). At the same time, though finding arbitrary values was (simply put) a rather arbitrary thing to do, the engineers-in-training at the École would have had many occasions to find roots in the course of their work and studies.

On the other hand, Cauchy leaves several potentially important ideas between the lines. We know, for instance, what it means for terms to be pairwise of opposite signs, but what does this mean in the limit of a sequence? Proofs today typically specify one sequence of values as approaching the intermediate value from below and the other from above. Indeed, it is striking that, though the proof discusses sequences and values with opposite signs, the only symbolic inequalities in the entire argument are used for values of x and never for values of $f(x)$. For both x and $f(x)$, Cauchy refers frequently not just to values but to *quantities*, implying that they have geometric magnitudes.

This explains, in part, why Cauchy so freely makes claims about the convergence and limits of the sequences obtained in his proof. As a pure matter of algebraic abstractions, one needs a lemma to assert that the common limit of the sequences $f(x_0), f(x_1), \ldots$ and $f(X), f(X'), \ldots$ would equal $f(a)$. As a matter of geometry, however, the identity might not strike one as requiring a separate argument or citation.

It would not be until several decades after Cauchy's course was published that mathematicians would systematically attempt to define teratological functions that defied the intuitions associated with the usual mechanical problems of polytechnical mathematics. In Cauchy's time, mathematical analysis remained first and foremost the mathematical study of the world—a world filled with complex phenomena but also a world exhibiting profound regularities. Cauchy's course, for instance, assumes that all continuous functions are differentiable; indeed, all the continuous functions he might care to differentiate were more or less smooth. In a sense, then, Cauchy's preference for geometry reflected a desire to remain true to the world and to use only those mathematical techniques that genuinely reflected worldly magnitudes, even if this limited what he could say mathematically about that world.[6]

Stuck in the Middle

The peculiarities of Cauchy's proofs help us see that the rigor Cauchy prized was something quite different from the rigor we now associate with his name. When Cauchy objected to the mathematics of his predecessors, he did not find them lacking in their adherence to formal rules for symbolic manipulation. Quite the contrary, he felt that mathematicians in the preceding century trusted such rules altogether too much. With this in mind, it is not surprising that Cauchy's project of reform was not, at its heart, based on carefully placed quantifiers, deftly manipulated sequences and inequalities, and meticulous logical exactitude.

To tame the dangerous fashion for algebra, Cauchy demanded a return to geometry in both its senses. He is remembered today for making his proofs systematic and logical, but his own proofs place a clearer emphasis on the geometry of magnitudes, not the geometry of methods. Cauchy sought to save mathematical analysis by ensuring that its powerful algebraic tools stayed true to the world of geometry, hence his insistence on convergence and his caution when defining imaginary and even negative numbers. Where he did not see any danger of symbols losing their geometric referents, as in his proofs of the intermediate value theorem, he could in fact be quite lax with their use.

In this sense, Cauchy's analysis appears surprisingly regressive. The rigor he advocated was a return to geometric reasoning that a century of mathematicians had rejected as stale, tedious, and counterproductive. His methods were difficult, and his students and colleagues frequently lamented their cumbersome impracticality. And yet, Cauchy seems now to have won the day.

How could such a reactionary mathematician so transform the mathematics of his generation in a way that now appears progressive and visionary? Cauchy realized that he could not do away with algebra even in his own mathematics. Rather than throw algebra out entirely, he worked to endow algebra with the virtuous rigors of geometry by developing algebraic criteria to match geometric reasoning. In so doing, he advanced the idea that it was possible to have it both ways: to enjoy the power of algebraic thinking while still adhering to the discipline and certitude of geometry. All one needed was to put the demands of geometry in algebraic terms.

Thus, Cauchy's rules for convergence and his attention to the limits of formal expressions' validity created new problems and new opportunities. He opened up ways of studying mathematical phenomena that remain vital nearly two centuries later. While his approach proved durable, its initial motivation could easily be forgotten. The ensuing generation of European mathematicians latched on to his disciplined way of studying the meanings of formal expressions while jettisoning his preoccupation with the geometry of magnitudes. From them, we have the beginnings of the set-theoretic foundations of analysis that undergraduates learn today.

Of course, not even the set theorists had the final word on rigor. Mathematicians must constantly balance what methods are worthwhile, what arguments are convincing, and what values are worth conveying to students. Those dismayed by the lack of consensus on these points today or wishing that opponents could simply see that their positions are illogical, unrigorous, or counterproductive can take comfort in the fact that these debates are not just a normal part of the history of mathematics, but that they can help to spur new ideas and approaches, often in unexpected ways. Studying the history of mathematics, we can appreciate that in some ways we are all, like Cauchy, stuck in the middle between our discipline's past and its open-ended future.

Notes

1. On Cauchy's interconnected politics, religion, and pedagogy, see [2].
2. When Cauchy wrote his textbook, non-Euclidean geometries were still just over the horizon of mathematical theory.
3. The specific body of theory and techniques we now call abstract algebra was, like non-Euclidean geometry, still just beginning to emerge as Cauchy wrote. Some of Cauchy's earliest work was on problems we might now consider in this area.
4. Cauchy's definition of continuity may also appear unusual to those expecting epsilons, deltas, quantifiers, and convergence. He defines continuous functions as those for which the difference $f(x + \alpha) - f(x)$ is infinitesimally small when α is [4, pp. 34–35]. There is considerable secondary literature on Cauchy's "infinitesimally small quantities" and their relationship to various ideas about variables, continuity, and convergence. In particular, although some have argued that his notions were ultimately equivalent to ideas developed in either "epsilontic" or nonstandard analysis, most agree that Cauchy omitted or left implicit many important ideas and intuitions about his infinitesimally small quantities.
5. One might be tempted to dismiss this proof as merely a pedagogically oriented plausibility argument, but Cauchy himself makes no such excuse for it, and such a move would be exceptional in a textbook meant to showcase his model of rigor. His allusion to the "direct and purely analytic" proof in the appendix is sometimes read as an admission that the above proof is inadequate, but this view substitutes later values of analytic rigor where Cauchy's own priorities are at best unclear. The best evidence of Cauchy's view remains the fact that he calls this argument a proof (something he does not do for every argument following a stated theorem) and places it prominently in the body of his textbook.
6. I elaborate in [1] how this mathematical impulse relates to Cauchy's religious and political conservatism.

References

[1] M. J. Barany. (2011). "God, king, and geometry: Revisiting the introduction to Cauchy's Cours d'analyse." *Historia Mathematica* 38(3), 368–88.
[2] B. Belhoste. (1991). *Augustin-Louis Cauchy: A Biography*, translated by F. Ragland. Springer, New York.

[3] R. E. Bradley and C. E. Sandifer. (2009). *Cauchy's Cours d'Analyse: An Annotated Translation*, Springer, New York.

[4] A.-L. Cauchy. (1821). *Cours d'Analyse de l'École Royale Polytechnique; 1ʳᵉ Partie, Analyse Algébrique*, Debure, Paris.

[5] J. V. Grabiner. (1974). "Is mathematical truth time-dependent?" *The American Mathematical Monthly* 81(4), 354–65.

[6] J. V. Grabiner. (1981).*The Origins of Cauchy's Rigorous Calculus*, MIT Press, Cambridge, MA.

[7] J. F. Montucla. (1802). *Histoire des mathématiques, nouvelle édition, considérablement augmentée, et prolongée jusque vers l'époque actuelle*, Vol. III, Henri Agasse, Paris.

Plato, Poincaré, and the Enchanted Dodecahedron: Is the Universe Shaped Like the Poincaré Homology Sphere?

Lawrence Brenton

A staple of fantasy/horror stories goes like this. You are imprisoned in an enchanted room. You try to escape by walking out the door. But as you step out through the door—you step in through the window! Is that weird, or what?! Could the universe be like that?

Well, I am not sure about the whole universe, but the surface of the Earth is certainly "like that."

In Figure 1A, suppose we want to escape the cruel world by exiting through the door stage right. In real life, when we step off the eastern edge of the world, we are not surprised to find ourselves slogging along in Alaska in the far West. We have gotten used to the fact that the world is shaped the way it is. Indeed, the only thing weird is how we drew the map.

Everything Old Is New Again

The job of physics is to determine the elementary constituents of the material world and to describe the forces that govern their behavior. The Hellenistic Greeks had it all figured out. There are four elements: earth, water, air, and fire. Each of these elements embodies, to a greater or lesser degree, either the force of gravity (earth, water, and air) or its opposite, the force of levity (fire).

This was quite a satisfactory classification because there are also four elemental shapes, the regular polyhedra (Platonic solids). The element earth is cubical, which is why we can build things by stacking soil and stone. Water is shaped like little round icosahedra, the better to roll

Trivia question: In what state is the easternmost point of the United States? Answer: Alaska. The longitude of Semisopochnoi Island, Alaska is 179° 37′ *East*.

FIGURE 1A. A closed two-dimensional surface. What happens when we exit through the door on the right?

FIGURE 1B. Are Siberia and Alaska far apart or close together? It all depends on who is drawing the map. Could the universe have the same property?

downhill. Air takes its shape from the bright and breezy octahedron. But don't stick your hand in the fire! The sharp points of myriad tiny tetrahedra will prick you.

There was only one problem. There are *five* regular polyhedra. What unearthly "quint-essence" does the fifth elemental shape, the dodeca-hedron, champion?

Plato, in the dialogue *Timaeus*, overcame this awkward dilemma by assigning the dodecahedron to "the Cosmos." Plato's prize stu-dent Aristotle proposed instead a new and noble substance, "ether," that ruled the superlunary realms. Since then, various forms of ether

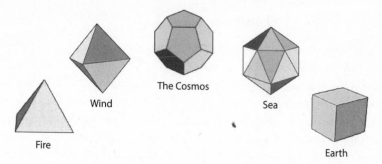

FIGURE 2. The five elements, then and now.

Classical Antiquity			Modern Particle Physics	
Element	Force	Polyhedron	Basic Form of Matter	Gravity/ Levity
Fire	Levity	Tetrahedron	Bosons (example: photon)	Gravity
Air	Gravity	Octahedron	Leptons (example: electron)	Gravity
Water	Gravity	Icosahedron	Dark matter	Gravity
Earth	Gravity	Cube	Quarks	Gravity
Cosmos/ ether/ quintessence	None	Dodecahedron	Dark energy/ cosmological constant	Levity

have periodically been invoked to account for phenomena that we did not fully understand. Its nineteenth-century incarnation was the "lumeniferous ether" (the stuff that does the waving when light waves propagate). This concept of quintessence was dispatched in 1887 by the famous Michelson-Morley experiment—only to morph into a new shape and rise again 111 years later as the vacuum energy of space ("dark energy").

But this story is about Plato. Plato thought that the universe is shaped like a dodecahedron. Was he right?

The Music of the Spheres

The Greeks were not the only ones who were intrigued by the symmetry of these shapes. Johannes Kepler, from 1596 to 1619, published several works that were destined to have a seismic effect on the scientific world. By developing his three laws of planetary motion, Kepler gave scientific substance and mathematical precision to the mind-boggling and counterintuitive speculation of Nicolaus Copernicus that we are not the center of the universe after all. Rather, the Earth is just another run-of-the-mill planet circling the master sun. The age of modern mathematical astronomy was born when Isaac Newton was able to give a derivation of Kepler's laws from his inverse square law of gravity.

But Kepler also had a fourth law, one that did not fare as well as the other three. Kepler noticed that the orbits of the six planets fit perfectly into the alternating inscribed and circumscribed spheres of the five Platonic solids nested one within the next. Surely, the creator of the universe would not be so cruel as to give us exactly six planets and exactly five Platonic solids that fit together so cleverly just to trick us by a divine hoax!

But hoax it turned out to be. Two centuries later, William and Caroline Herschel discovered a seventh planet, George. (George's Star, so named to honor the Herschels' patron, King George III of England, was later renamed Uranus.) And in any case, careful measurements, even in Kepler's time, revealed that the ratios of the planetary orbits were not quite as perfect as Kepler had at first hoped. In the end, the scientist won out over the mystic, and Kepler reluctantly abandoned this geometric model of the solar system.

Although Kepler's "Fourth Law" did not work out, his Second Law did. This law predicts that a planet, in its elliptical journey around the sun, goes slightly faster when it is closer to the sun and slightly slower when it is farther away. For the planet Earth, the ratio of the two extreme angular velocities is very close to the ratio of the frequencies of the notes "fa" and "mi" on the musical scale. Thus, the Earth's contribution to the eternal music of the spheres is the doleful dirge "fa, mi, fa, mi, fa, mi, ..." According to Kepler, this stands for "famine, misery, famine, misery . . . ," which captures the human condition on Earth and accounts for our longing for heaven. And so we turn at last to the imponderably vast reaches of space that lie beyond our tiny local system of

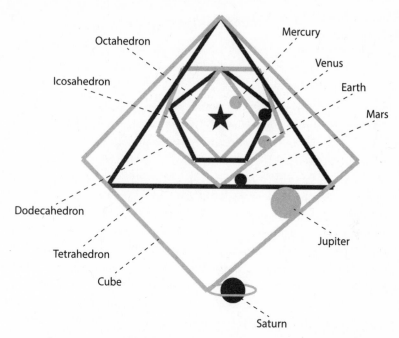

Kepler's ratios for planetary orbits. Are these ratios close enough to warrant further investigation of Kepler's intriguing speculation?

FIGURE 3. Kepler's vision of the solar system. The orbit of each successive planet lies on the sphere circumscribed about a regular polyhedron and inscribed within the next.

Polyhedron (side length = 2)	Inner Radius	Outer Radius	Ratio	Planet	Distance from Sun*	Ratio
				Mercury	36	1.9
Octahedron	$\frac{2}{3}$	$\sqrt{2}$	2.1	Venus	67	
						1.4
Icosahedron	$\frac{4\cos^2 36°}{\sqrt{3}}$	$2\sin 72°$	1.3	Earth	93	
						1.5
Dodecahedron	$\frac{2\cos^2 36°}{\sin 36°}$	$2\sqrt{3}\cos 36°$	1.3	Mars	142	
						3.4
Tetrahedron	$\frac{1}{\sqrt{6}}$	$\sqrt{\frac{3}{2}}$	3.0	Jupiter	484	
						1.8
Cube	1	$\sqrt{3}$	1.7	Saturn	887	

*average, in millions of miles

planets. What dare we hope to say about the geometric features of the cosmos in its entirety?

Is the Universe Finite or Infinite?

Confronted by this perplexing riddle, our imagination is sorely tested whichever way we jump. If the universe is finite, then what's "out there" beyond the finite universe? Albert Einstein popularized the language "finite but unbounded." The surface of the Earth has finite extent. Yet we can march forward forever in a straight line (that is, as straight as possible, following the curvature of the Earth) and never come to the end of the world. Compact models of the three-dimensional spatial universe have the same property. Although of finite size, there is no "beyond." A spaceship can set out in a straight line and travel forever, perhaps visiting the same regions of space over and over, but never exit the universe nor come to a boundary barring its way.

Conversely, suppose that the universe is infinite. In our imagination, let's divide up the infinite universe into infinitely many identical regions, each as large as we like. The visible universe contains about 10^{80} atoms. That's a lot, but it is still a finite number. There are only finitely many ways to arrange a finite number of particles with respect to relative locations and velocities. If the universe is truly infinite, then probabilistic considerations suggest that everything that can possibly happen *has* happened and, indeed, is happening right now in infinitely many different parts of the infinite universe. Compared to this scenario, is it so far-fetched to imagine that the universe has a finite size and a "shape"? But what could the shape possibly be?

In 2003, a group of astronomers led by Jean-Pierre Luminet at the Observatory of Paris announced that studies of the pattern of hot spots in the cosmic microwave background radiation give evidence that the entire universe is shaped like—what else?—a dodecahedron! Alas for Plato, I must add that more recent and extensive studies of the cosmic background radiation since 2003 have not given encouragement to this conclusion, and the dodecahedral model is probably wrong. Nevertheless, an examination of this model is useful as a guide to what we should be looking for as we attempt to determine the shape and size of the whole shebang.

The Poincaré Conjecture and the Poincaré Homology Sphere

In 1904, Henri Poincaré, a pioneer of the discipline of algebraic topology, posed a conundrum that baffled the mathematical world for almost a century. This was the famous Poincaré conjecture. It was not settled until 2002, by the reclusive Russian mathematician Grigori Perelman. The Clay Mathematics Institute had offered a million-dollar prize for a solution to Poincaré's problem, but Perelman refused to accept the money. He felt that the contributions of other researchers—especially of Richard Hamilton at Columbia University—were just as important to the eventual solution as were his own.

The Poincaré conjecture states that if a three-dimensional continuum (for example, the universe) is closed, without boundary, and simply connected (every closed path can be contracted to a point), then it is topologically equivalent to the three-dimensional sphere. In other words, if we find a crumpled-up three-dimensional bit of rubber (while wandering about in four-dimensional space) with the stated properties, then we can blow it up into a big, round balloon. The blowing-up process that turned out to be successful is called the Ricci flow. This is a technique in differential geometry for continuously deforming the object, making it progressively rounder and rounder.

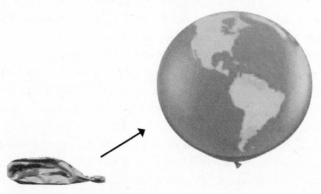

FIGURE 4. The Poincaré conjecture (now the Poincaré-Hamilton-Perelman theorem) states that every simply connected compact three-dimensional manifold is homeomorphic to the three-sphere. The two-dimensional analogue is pictured here.

The reason we care about this property of simple connectivity in cosmology is as follows: Suppose we go on a long journey through space and end up where we started, pausing periodically along the way to solve a differential equation that captures some law of nature (conservation of energy, for instance). If space is simply connected, then all these local solutions are guaranteed to fit together and give a global solution, so that when we get back home we find that the properties of the physical world are the same as when we left.

But this characteristic (technically called *vanishing de Rham cohomology*) is weaker than simple connectivity. A three-dimensional space could have this desirable conservation property even if it is not simply connected. It is sufficient that the space be what is called a *homology sphere*. Indeed, Poincaré's original conjecture was that every three-dimensional homology sphere is topologically equivalent to the standard three-sphere. But Poincaré quickly constructed a counterexample to his own conjecture, which led him to turn his attention to homotopy theory and simple connectivity.

Poincaré's construction of the dodecahedral homology sphere is rather technical. But in 1929, topologist Hellmuth Kneser (with further elucidations by Herbert Seifert and William Threlfall in 1931) produced a model of this object that is easy to describe. Start with an ordinary solid dodecahedron. With all due respect to Plato, this cannot be a model for the universe because it has a boundary, comprising the 12 pentagonal faces. If we lived in the interior of such a universe, we could journey outward until we came to one of the 12 walls. Now what?

In real life, there is no wall at the end of the universe. Since we cannot escape the universe no matter what we do, when we pass through the side of the dodecahedron, somehow or other we must end up back in the same old universe. We exit through a door only to come back in through a window.

Mathematically, the construction is to take each pentagonal face of the dodecahedron, rotate it through a tenth of a circle, and identify it with the face opposite. When we go out a door, we come in a window rotated through 36 degrees. This construction provides a model of the cosmos that is viable, and one with which Plato would have been pleased.

Figure 5 provides a three-dimensional map of the Poincaré homology sphere. In Figure 5A, we see an imaginary view of the enchanted dodecahedron from the outside, with six of the 12 walls visible. If we

FIGURE 5. The enchanted dodecahedron. You can't check out of the Plato-Poincaré hotel.

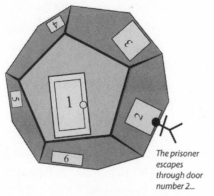

FIGURE 5A. Imagined outside view.

The prisoner escapes through door number 2...

...to no avail. He reenters through window number 2

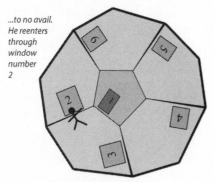

FIGURE 5B. Cutaway view of the inside.

FIGURE 5C. A different view. Door number 2 and window number 2 are actually one and the same. (Compare to Figure 1B.)

Figures 5A and 5B together comprise a three-dimensional map of the dodecahedral universe. Just as Siberia and Alaska appear to be far apart in the map of the world shown in Figure 1A, so door number 2 and window number 2 appear to be far apart in this particular rendering of the map.

attempt to exit through one of these doors, we simultaneously reenter through the window opposite, as shown in the cutaway view of the interior, Figure 5B.

Historically, terrestrial mapmakers quickly learned that they could not draw a completely satisfactory map of the entire spherical Earth on a flat two-dimensional piece of paper. By the same token, we cannot display the true shape of the cosmos by constructing models in ordinary three-dimensional space. We are handicapped in that we cannot draw pictures in four dimensions, and we cannot exit the universe to examine it from an outside vantage point. So we must do the best we can by reading our three-dimensional maps.

In Figures 5A and 5B, door number 2 and window number 2 appear to be far apart. This is a feature of the particular way that we drew the map, and a misleading one at that. In reality, door number 2 and window number 2 are immediately adjacent, as shown in Figure 5C. Indeed, the doors and windows are not there at all, any more than there is a physical barrier between Siberia and Alaska.

No A-Geometers Needed

According to tradition, visitors to Plato's Academy, founded in 387 BCE, were greeted by the inscription *Μηδείσ αγεωμέτρητος εισίτωμοι την θύραν*. Which is to say, "Let no one ignorant of geometry enter upon my door." A few decades later, Euclid of Alexandria, perhaps himself a student at the academy at one time, undertook the compilation of what was destined to become the number one best-selling secular book of all time. *The Elements* has been issued in hundreds of editions and has never been out of print since the invention of printing. By setting forth the logical foundations of geometric reasoning, Euclid provided a model for all subsequent mathematical argumentation and established a standard of truth that other branches of learning can only envy.

But by Poincaré's time (the end of the nineteenth century), investigators such as János Bolyai, Nikolai Lobachevsky, Eugenio Beltrami, and Poincaré himself had elevated non-Euclidean geometries—those axiomatic systems that deny Euclid's parallel line postulate—to the same logical status as Euclidean geometry. If Euclidean geometry is logically consistent, then so is non-Euclidean geometry. Indeed, Poincaré believed that the only reason we tend to see the world through Euclidean

FIGURE 6A. The view from inside the Poincaré homology sphere. No matter how many doors we pass through, we are still on the inside.

FIGURE 6B. Are we visiting many different rooms or the same room over and over?

eyes is because of natural selection among our prehuman ancestors: "Geometry is not true, it is advantageous."

The geometry of the Poincaré homology sphere is non-Euclidean, three-dimensional, spherical geometry. If we draw a huge triangle in space, the sum of the angles is a bit greater than 180 degrees. The circumference of a circle is slightly less than $2\pi r$. And "parallel" lines eventually meet. In principle, such geometric properties can be tested in the real universe by observation and experiment. So far, all observed cosmological data are consistent with the Euclidean model. Perhaps Euclid will have the last laugh on Plato and Poincaré after all.

In the meantime, all we can say with confidence is that if the universe is an enchanted dodecahedron, it is a really, really big one.

Note

This paper is embellished and enlivened by a series of flash animations created by Dimitra Tzianni, available at http://www.math.wayne.edu/~brenton/flash. The text is complete without the animations, but not nearly so whimsical.

Further Reading

Jorge Luis Borges, *The Library of Babel*. http://jubal.westnet.com/hyperdiscordia/library_of
_babel.html.

Jean-Pierre Luminet, Glenn D. Starkman, and Jeffrey Weeks, "Is Space Finite?" *Scientific American* (April 1999), 90–97. http://cosmos.phy.tufts.edu/~zirbel/ast21/sciam/IsSpace
Finite.pdf.

Plato, *Timaeus*, translated by Benjamin Jowett, Internet Classics. http://classics.mit.edu/Plato
/timaeus.html.

Computing with Real Numbers, from Archimedes to Turing and Beyond

MARK BRAVERMAN

We are so immersed in numbers in our daily lives that it is difficult to imagine that humans once got by without them. When numbers were finally introduced in ancient times, they were used to represent specific quantities (such as commodities, land, and time); for example, "four apples" is just a convenient way to rephrase "an apple and an apple and an apple and an apple"; that is, numbers had algorithmic meaning millennia before computers and algorithmic thinking became as pervasive as they are today. The natural numbers 1, 2, 3, . . . are the easiest to define and "algorithmize." Given enough time (and apples), one can easily produce a pile with any natural number of apples.

Fractions are not as easy to produce as whole numbers, yet the algorithm for them is fairly straightforward. To produce 2/3 of an apple, one can slice an apple into three equal parts and then take two of them. If one considers positive rational numbers, there is little divergence between the symbolic representation of the number and the algorithm one needs to "construct" this number out of apples; the number practically shouts a way to construct it. These numbers were the ones that populated the world of the ancient Greeks (in Archimedes' time), who often viewed numbers and fractions through the lens of geometry, identifying them with geometric quantities. In geometric language, natural numbers are just integer multiples of the unit interval, and positive rational numbers are integer fractions of these intervals.

It was tempting at the time to believe that all numbers, or all possible interval lengths, are rational and can be constructed in this manner. However, it turns out not to be the case. The simplest example of an irrational number is $\sqrt{2}$. The number $\sqrt{2}$ is easily constructed geometrically (such as by using a ruler and compass) and is the length of the diagonal of a 1×1 square. On the other hand, a simple elegant proof, first given by the Pythagorean philosopher Hippasus, shows that one cannot write $\sqrt{2}$ as m/n for integers m and n. Hippasus' result was controversial at the time since it violated the belief that all mathematical quantities are rational. Legend has it that Hippasus was drowned for his theorem. History offers many examples of scientists and philosophers who suffered for their discoveries, but we are not aware of another example of a mathematician being punished for *proving* a theorem. In modern terms, the conflict can be framed through a question: What family of algorithms suffices if one wants to compute all real numbers representing plottable lengths? Hippasus' opponents supposed integer division would suffice.

Even more intriguing is the number π, which represents the circumference of a circle of diameter 1. Perhaps the most prominent mathematical constant, π can also be shown to be irrational, although the proof is not as simple as for $\sqrt{2}$ and was not known in ancient times. We cannot represent either $\sqrt{2}$ or π as rational fractions. We can "construct" them from mesh wire using their geometrical interpretations, but can we also figure out their numerical values? Unlike the names 4 and 2/3, the names $\sqrt{2}$ and π are not helpful for actually evaluating the numbers. We can calculate approximations of these numbers; for example, for π, we can write

$$3.1415926 < \pi < 3.1415927$$

or perhaps we can follow Archimedes, who carried out the earliest theoretical calculations of π, and write

$$\frac{223}{71} < \pi < \frac{22}{7}.$$

Both representations are correct, giving us a good handle on the value of π, but both have limited precision, thus losing some information about the true value of π. All real numbers can be written using their infinite binary (or decimal) expansion, which can be used to name the number, specifying it unambiguously.

The infinite representation $\pi = 3.1415926\ldots$ unambiguously specifies the number π. But, alas, we cannot use it to write π in a finite amount of space. An ultimate representation would take a finite amount of space but also allow us to compute π with any desired precision.

In modern language, such a representation should be algorithmic. As there are many formulas for π, there are likewise many ways to represent π this way; for example, in approximately the year 1400, Madhava of Sangamagrama gave this formula:

$$\pi = 4\sum_{k=0}^{\infty} \frac{(-1)^k}{2k+1} = \frac{4}{1} - \frac{4}{3} + \frac{4}{5} - \frac{4}{7} + \ldots \tag{1}$$

It allows us to compute π with any precision, although the convergence is painfully slow; to compute the first n digits of π, one needs to take approximately 10^n terms of this sum. Many elegant formulas for computing π have been devised since, some allowing us to compute π in time polynomial in the number of digits. One such formula, known as the Bailey-Borwein-Plouffe formula, is given by

$$\pi = 4\sum_{k=0}^{\infty}$$

$$\left[\frac{1}{16^k}\left(\frac{4}{8k+1} - \frac{2}{8k+4} - \frac{1}{8k+5} - \frac{1}{8k+6} \right) \right]. \tag{2}$$

The fact that the terms in formula (2) decrease exponentially fast in k causes the sum to converge rapidly. Mathematically speaking, formulas (1) and (2) are both valid "names" for π, although the latter is better because it corresponds to a much more efficient algorithm.

Can all numbers be given names in a way that allows us to compute them? No, as it turns out. Surprisingly, it took until Alan Turing's seminal paper in 1936 (Turing 1936) to properly pose and answer the question. Turing had to overcome a profound philosophical difficulty. When showing that a real number is computable, we would need only to describe an algorithm able to compute it with any prescribed precision, as we did with the number π. In showing that a number $x \in \mathbb{R}$ is not computable, we need to rule out all potential ways of computing x. The first major step in any such proof is formalizing what "computing" means by devising a model of computation. This is exactly what Turing did, defining his famous Turing Machine as an abstract device capable of performing all mechanical computations. Turing's paper started the modern field of computability theory. Remarkably, it happened about a dozen years before the first computers (in the modern sense of the word) were built. Turing used his new theory to define the notion of computable numbers. Not surprisingly, a modern reinterpretation of

Turing's definition says that a number x is computable if we can write a C++ or Java program that (given sufficient time and memory) can produce arbitrarily precise approximations of x. One of Turing's key insights was the Halting Problem $\mathcal{H}$ (which takes an integer n and outputs $\mathcal{H}(n) = 1$ if and only if $n = [P]$ is an encoding of a valid program P and P terminates) is "undecidable"; no algorithm exists that, given a program P, is capable of deciding whether or not P terminates.

The Halting Problem allows us to give a specific example of a noncomputable number. Write down the values of the function $H(\bullet)$; the number

$$X_{\mathcal{H}} = 0.\mathcal{H}(1)\mathcal{H}(2)\mathcal{H}(3)\ldots = \sum_{n=1}^{\infty} 10^{-n}\mathcal{H}(n).$$

is not computable, since computing it is equivalent to solving the Halting Problem. Fortunately, "interesting" mathematical constants (such as π and e) are usually computable.

One reason numbers and mathematics were developed in the first place was to understand and control natural systems. Using the computational lens, we can rephrase this goal as reverse-engineering nature's algorithms. Which natural processes can be computationally predicted? Much of this article is motivated by this question. Note, unlike digital computers, many natural systems are best modeled using continuous quantities; that is, to discuss the computability of natural systems, we have to extend the discrete model of computation to functions and sets over the real numbers.

Real Functions and Computation

We have established that a number $x \in \mathbb{R}$ is computable if there is an algorithm that can compute x with any prescribed precision. To be more concrete, we say an algorithm $\mathcal{A}_x$ computes x if, on an integer input $n \in \mathbb{N}$, $\mathcal{A}_x(n)$ outputs a rational number x_n such that $|x_n - x| < 2^{-n}$. The algorithm $\mathcal{A}_x$ can be viewed as a "name" for x, in that it specifies the number x unambiguously. The infinite-digit representation of x is also its "name," albeit not compactly presented.

What does it mean for a function $f:\mathbb{R} \to \mathbb{R}$ to be computable? This question was first posed by Banach, Mazur, and colleagues in the Polish school of mathematics shortly after Turing published his original paper, starting the branch of computability theory known today as *computable analysis*. Now step back to consider discrete Boolean functions. A Boolean function $F:\{0,1\}^* \to \{0,1\}^*$ is computable if there is a program $\mathcal{A}_F$, that given a binary string $s \in \{0,1\}^*$, outputs $\mathcal{A}_F(s) = F(s)$. By analogy, an algorithm computing a real-valued function f would take a real number x as input and produce $f(x)$ as output. Unlike the Boolean case, input and output must be qualified in this context. What we would like to say is given a name for the value $x \in \mathbb{R}$, we should be able to produce a name for the output $f(x)$; that is, we want $\mathcal{A}_f$ to be a program that, given access to arbitrarily good approximations of x, produces arbitrarily good approximations of $f(x)$.

> A function $f:(a,b) \to \mathbb{R}$ is computable if there is a discrete algorithm $\mathcal{A}_f$ that, given a precision parameter n and access to arbitrarily good rational approximations of an arbitrary input $x \in (a,b)$, outputs a rational y_n such that
>
> $$|y_n - f(x)| < 2^{-n}.$$

This definition easily extends to functions that take more than one input (such as the arithmetic operations $+: \mathbb{R} \times \mathbb{R} \to \mathbb{R}$ and $\times: \mathbb{R} \times \mathbb{R} \to \mathbb{R}$). As with numbers, all "nice" functions, including those usually found on a scientific calculator, are generally computable. Consider the simple example of the function $f(x) = x^2$ on the interval $(0,1)$. Our algorithm for squaring numbers should be able to produce a 2^{-n} approximation of x^2 from approximations of x. And consider this simple algorithm

```
SimpleSquare(x,n)
```

   1. Request $q = x_{n+1}$, a rational $2^{-(n+1)}$-approximation of the input x.
   2. Output q^2.

Note that the algorithm `SimpleSquare` operates only with rational numbers. To see that the algorithm works, we need to show that for all $x \in (0,1)$ the output q^2 satisfies $|x^2 - q^2| < 2^{-n}$. Since x is in the interval $(0,1)$, without loss of generality we may assume that q is also in $(0,1)$. Therefore, $|x + q| < |x| + |q| < 2$, and we have

$$|x^2 - q^2| = |x + q| \cdot |x - q| \le 2\,|x - q| < 2 \cdot 2^{-(n+1)} = 2^{-n}.$$

This inequality shows that the `SimpleSquare` algorithm indeed produces a 2^{-n} approximation of x^2. Although the function $f(x) = x^2$ is computable on the entire real line $\mathbb{R} = (-\infty, \infty)$, in this case, the algorithm would have to be modified slightly to work.

A more interesting example is the function $g(x) = e^x$, which is defined on the entire real line. Indeed, for any x, we can compute e^x with any precision by requesting a sufficiently good rational approximation q of x and then using finitely many terms from the series

$$e^q = \sum_{n=0}^{\infty} \frac{q^n}{n!} = 1 + \frac{q}{1} + \frac{q^2}{2} + \frac{q^3}{6} + \frac{q^4}{24} + \dots$$

Throughout the discussion of the computability of these functions, we did not have to assume that the input x to a computable function is itself computable. As long as the Request command gives us good approximations of x, we do not care whether these approximations were obtained algorithmically. Now, if the number x is itself computable, then the Request commands may be replaced with a subroutine that computes x with the desired precision. Thus, if f is computable on (a,b) and $x \in (a,b)$ is a computable number, then $f(x)$ is also a computable number. In particular, since e^x is a computable function and π is a computable number, e^π and e^{e^π} are computable numbers as well.

One technical limitation of the Request-based definition is that we can never be sure about the exact value of the input x; for example, we are unable to decide whether or not the real-valued input x is equal to, say, 42. Thus the function

$$f(x) = \begin{cases} 1 & \text{if } x = 42.0 \\ 0 & \text{otherwise} \end{cases}$$

is not computable. The reason for this inability is that although we can Request x with any desired precision, no finite-precision approximation of x will ever allow us to be sure that x is exactly 42.0. If we take the requested precision high enough, we may learn that $x = 42.0 \pm 10^{-1000}$. This still does not mean that $f(x) = 1$ because it is possible that the first disagreement between x and 42.0 occurs after the 1000th decimal place (such as if $x = 42 + 2^{-2000} \neq 42.0$). The Request function can be viewed as a physical experiment measuring x. By measuring x, we can narrow down its value to a very small interval but can never be sure of its exact value. We refer to this difficulty as the "impossibility of exact computation." More generally, similar reasoning shows that only continuous functions may be computable in this model.

On the other hand, and not too surprisingly, all functions that can be computed on a calculator are computable under this definition of function computability. However, as with Turing's original work, the main goal of having a model of computation dealing with real functions is to tell us what *cannot* be done, or proving fundamental bounds on our ability to computationally tackle continuous systems. First we need to explore the theory of computation over the real numbers a little further.

COMPUTABILITY OF SUBSETS IN $\mathbb{R}^d$

In addition to numbers and functions, we are also interested in computing sets of real numbers; see Figure 1 for example subsets of $\mathbb{R}^2$. Sets in which we might be interested include simple geometric shapes (such as a circle), graphs of functions, and the more complicated ones, like the Koch snowflake and the Mandelbrot set. When is a set S in, say, the plane $\mathbb{R}^2$, computable? It is tempting to mimic the discrete case and say that S is computable whenever the membership function

$$x_S(x) = \begin{cases} 1 & \text{if } x \in S \\ 0 & \text{otherwise} \end{cases}$$

is decidable. However, this definition involves a serious technical problem: the same impossibility-of-exact-computation problem present in the $x \stackrel{?}{=} 42.0$ example. If x happens to lie on the boundary of S, we will never be able to decide whether $x \in S$ through a finite number of Request queries. To address this problem, we proceed by analogy

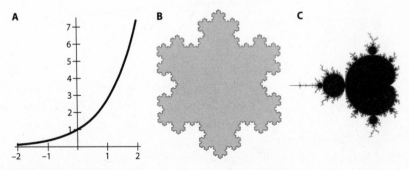

FIGURE 1. Examples of subsets of $\mathbb{R}^2$: the graph of $y = e^x$, the Koch snowflake, and the Mandelbrot set.

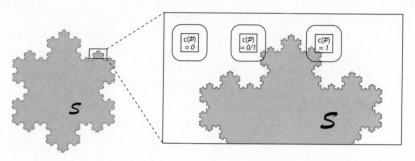

FIGURE 2. The process of deciding the color $c(P)$ for an individual pixel P.

with the computability of numbers. Rather than try to compute S, we should try to approximate it with any prescribed precision 2^{-n}.

What does it mean to "approximate" a set? There are many ways to address this question, and many "reasonable" definitions are equivalent. First take a "graphical" approach. Consider the process of producing a picture of the set S. The process would consist of many individual decisions concerning whether to color a pixel black or white; for example, we need to make 600×600 such decisions per square inch if S is being printed on a 600-dot-per-inch printer. Thus a discussion about drawing S with precision 2^{-n} can be reduced to a discussion about deciding on the color of individual pixels, bringing us back to the more familiar realm of 0/1-output algorithms.

To be concrete, let S be a subset of the plane $\mathbb{R}^2$ and let

$$\mathcal{P} = [x - 2^{-n-1}, x + 2^{-n-1}] \times [y - 2^{-n-1}, y + 2^{-n-1}]$$

be a square pixel of dimensions $2^{-n} \times 2^{-n}$. The coloring of pixels should satisfy the following conditions:

1. If $\mathcal{P}$ intersects with S, we must color it black;
2. If $\mathcal{P}$ is 2^{-n}-far from S, we must color it white. It is natural to ask why we don't simply require $\mathcal{P}$ to be colored white if it does not intersect S. The reason is, if we did, we would again run into the impossibility of exact computation. Thus we allow for a gray area in the image; and
3. If $\mathcal{P}$ does not intersect S but is 2^{-n} close to it, we do not care whether it is colored black or white.

FIGURE 3. One of the "easiest" (left) and one of the "hardest" (right) dynamical systems.

This gray area allows us to avoid the impossibility of the exact computation problem while still producing a faithful image of S (Figure 2).

The definition of set computability presented here may seem ad hoc, appearing to be tied in to the way we choose to render the set S. Somewhat surprisingly, the definition is robust—equivalent to the "mathematical" definition of S being "approximable" in the Hausdorff metric, a natural metric one can define on subsets of $\mathbb{R}^d$. The definition is also equivalent to the distance function $d_S(x)$, which measures how far the point x is from the set S, being a computable function.

Just as "nice" calculator functions are computable, "nice" sets are likewise computable; for example, a circle $C(o,r)$ with center $o = (x,y)$ and radius r is computable if and only if the numbers x, y, and r are computable. Graphs of computable functions are computable sets. Thus the graph of the function $x \mapsto e^x$ is computable. For a more interesting example, consider the Koch snowflake K (Figure 1B). This fractal set has dimension $\log_3 4$ and lacks a nice analytic description. However, it is computable and is, in fact, the limit set of a sequence of finite snowflakes. Each finite snowflake K_n is just a polygon and thus easily computable. To approximate K, we need to draw only the appropriate finite snowflake K_n, exactly how the Koch snowflake is drawn in practice, as in Figure 1B.

Now that we have the notion of computable real functions and real sets, we can turn to formulating the computational hardness of natural and artificial systems, as studied in the area of dynamical systems.

Computing Nature: Dynamical Systems

At a high level, the area of dynamical systems studies all systems that evolve over time. Systems ranging from an electron in a hydrogen atom, to the movement of galaxies, to brain activity can thus be framed in the context of dynamical systems. A dynamical system consists of a set of states $\mathcal{X}$ and a set of evolution rules $\mathcal{R}$. Evolution of the system occurs over time. The state of the system at time t is denoted by $X_t \in \mathcal{X}$. The time may move either discretely or continuously. If time is discrete, the evolution of the system is given by the sequence $X_1, X_2, X_3, \ldots$, and the rule $\mathcal{R}$ specifies the dependence of X_{t+1} on X_t. If time is continuous, the evolution $\mathcal{R}$ of the system $X_t = X(t)$ is usually given by a set of differential equations specifying the rates of change in $X(t)$ depending on its current state.

As an example, consider a simple harmonic oscillator. A mass in Figure 3 is attached to a spring and is moving in a periodic fashion. Assuming no friction, the system X_t^{osc} evolves in continuous time, and its state at any time is described fully by two numbers: the location of the mass on the line and its velocity. Thus the state X_t^{osc} can be represented by the vector $X_t^{osc} = (\ell(t), v(t))$, where $\ell(t)$ represents the location of the mass, and $v(t)$ represents its velocity. The evolution rule of the system is given in this case by high school physics:

$$\begin{cases} \ell'(t) = v(t) \\ v'(t) = -\alpha \cdot \ell(t) \end{cases}$$

where α is a parameter that depends on the spring and the mass of the weight. X^{osc} is a very simple system, and we can answer pretty much any question about it; for example, we can solve this system of equations to obtain the full description of the system's behavior

$$X_t^{osc} = (A \sin(\sqrt{\alpha} \cdot t + \phi), A\sqrt{\alpha} \cos(\sqrt{\alpha} \cdot t + \phi)) \qquad (4)$$

where the parameters A and ϕ depend on the initial condition $X_0^{osc} = (\ell(0), v(0))$ of the system at time 0; that is, if we know the exact state of the system at time 0, we can compute the state of the system at any time in the future. It is not just the prediction problem that is easy for this system. Using the analytic solution (4), we can answer almost any question imaginable about it; for example, we can describe the set of all possible states the system being released from state X_0^{osc} will reach.

At the other extreme, predicting some dynamical systems in the long run is incredibly difficult. One important set of examples of "hard" dynamical systems comes from computer science itself. Consider the Turing Machine or its modern counterpart, a RAM computer with unlimited RAM, as a dynamical system. The state-space X^{comp} is the (infinite) state space of the computer. The system X^{comp} evolves in discrete time, with X_t^{comp} representing the state of the computer at step t of the execution. The evolution rule $\mathcal{R}$ is the rule according to which the computation proceeds; that is, $\mathcal{R}(X)$ is the state of the computer in the next time step if its current state is X. Call this system the *computer dynamical system*.

The computer dynamical system is easy to simulate computationally; all we must do is simulate the execution of the computation. On the other hand, unlike the oscillator example, answering long-term questions about the system is difficult; for example, given an initial condition X_0^{comp}, there is no computational procedure that can tell us whether the system will ever reach a given state Y; determining whether the system will reach a terminating state is equivalent to solving the Halting Problem for programs, which, as discussed, is computationally undecidable. It can likewise be shown that almost any nontrivial question about the long-term behavior of X^{comp} is noncomputable in the worst case.

These examples exist at the two extremes of computational hardness: X^{osc} is a linear dynamical system and fully computable. X^{comp} is (trivially) computationally universal, capable of simulating a Turing Machine, and reasoning about its long-term properties is as computationally hard as solving the Halting Problem. What kinds of systems are prevalent in nature? For example, can an N-body system evolving according to the laws of Newtonian gravity simulate a computer, in which case predicting it would be as difficult as solving the Halting Problem? Is predicting it computationally easy? Or something in between?

We do not know the answers for most natural systems. Here, we consider an interesting in-between example that is one of the best-studied dynamical systems. We consider dynamics on the set of complex numbers $\mathbb{C}$, or a number of the form $a + bi$, evolving by a quadratic polynomial. For the rest of this discussion, the set of complex numbers is identified with the 2-D complex plane, with the number $a + bi$ corresponding to the point (a, b), allowing us to visualize subsets of $\mathbb{C}$ nicely. Let $c \in \mathbb{C}$ be any complex number. Denote

$$p_c(z) := z^2 + c.$$

Define the discrete-time dynamical system X_t^c by

$$X_{t+1}^c = p_c(X_t^c).$$

The polynomial $p_c(z)$ is arguably the simplest nonlinear transformation one can apply to complex numbers, yet this system is already complicated enough to exhibit a variety of interesting and complicated behaviors. In particular, it is impossible to give a closed-form expression for X_t^c in terms of X_0^c, as we did with the oscillator example. Dynamical systems of the form X_t^c are studied by a branch of the theory of dynamical systems known as complex, or holomorphic, dynamics. Within mathematics, one of the main reasons for studying these systems is that the rich variety of behaviors they exhibit allows us to learn about the behavior of more general (and much more difficult to study) dynamical systems.

Outside mathematics, complex dynamics is best known for the fascinating fractal images it generates. These images, known as Julia sets (Figure 4), depict a global picture relevant to the long-term behavior of the system X^c. More specifically, J_c is the subset of initial conditions in the complex plane on which the long-term behavior of X^c is unstable. To understand what this means, we need to take a slightly closer look at the system X^c. Consider an initial point $X_0^c = x_0$, as mapped by $p_c(z) = z^2 + c$:

$$x_0 \mapsto x_0^2 + c \mapsto (x_0^2 + c)^2 + c \mapsto \ldots$$

If we start with an x_0 with a very high absolute value, say, $|x_0| > |c| + 2$, then the absolute value of $p_c(x_0) = x_0^2 + c$ will be larger than $|x_0|$, $|p_c(p_c(x_0))|$ will be larger still, and the state of the system will diverge to ∞. The set of starting points for which the system does not diverge to

FIGURE 4. Example Julia sets.

∞ is called the *filled Julia set* of the system X^c and is denoted by K_c. The Julia set[1] J_c is the boundary ∂K_c of the filled Julia set.

The Julia set J_c is the set of points around which the system's long-term behavior is highly unstable. Around each point z in J_c are points (just outside K_c) with trajectories that ultimately escape to ∞. There are also points (just inside K_c) with trajectories that always stay within the bounded region K_c. The Julia set itself is invariant under the mapping $z \mapsto z^2 + c$. This means that trajectories of points that start in J_c stay in J_c.

The Julia set J_c provides a description of the long-term properties of the system X_c. Julia sets are therefore valuable for studying and understanding these systems. In addition, as in Figure 4, Julia sets give rise to an amazing variety of beautiful fractal images. Popularized by Benoit Mandelbrot and others, Julia sets are today some of the most drawn objects in mathematics, and hundreds of programs for generating them can be found online.

Formally speaking, the problem of computing the Julia set J_c is one of evaluating the function $\mathcal{J}: c \mapsto J_c$. $\mathcal{J}$ is a set-valued function whose computability combines features of function and set computability discussed earlier. The complex input $c \in \mathbb{C}$ is provided to the program through the request command, and the program computing $\mathcal{J}(c) = J_c$ is required to output an image of the Julia set J_c within a prescribed precision 2^{-n}. $\mathcal{J}$ is a fascinating function worth considering in its own right; for example, the famous Mandelbrot $\mathcal{M}$ (Figure 1C) can be defined as the set of parameters c for which $\mathcal{J}(c)$ is a connected set. It turns out that the function $\mathcal{J}(c)$ is discontinuous, at least for some values of c (such as for $c = 1/4 + 0 \cdot i$). This means that for every ε there is a parameter $c' \in \mathbb{R}$ that is ε-close to $1/4$ but for which the Julia set $J_{c'}$ is very far from $J_{1/4}$. Because of the impossibility of exact computation, this discontinuity implies that there is no hope of producing a single program $P_{\mathcal{J}}$ that computes $\mathcal{J}(c)$ for all parameters c; on inputs close to $c = 1/4$, such a program would need to use request commands to determine whether c is in fact equal to $1/4$ or merely very close to it, which is impossible to do.

If there is no hope of computing $\mathcal{J}$ by one program, we can at least hope that for each c we can construct a special program P_{J_c} that evaluates J_c. Such a program would still need access to the parameter c through the request command, since only a finite amount of information about the continuous parameter $c \in \mathbb{C}$ can be "hardwired" into the program P_{J_c}. Nonetheless, by requiring P_{J_c} to work correctly on only one input c, we manage to sidestep the impossibility of the exact computation problem.

Most of the hundreds of online programs that draw Julia sets usually draw the complement $\bar{K}_c = \mathbb{C} \setminus K_c$ of the filled Julia set, or the set of points whose trajectories escape to ∞. It turns out that from the computability viewpoint, the computability of $\bar{K}_c$ is equivalent to the computability of J_c, allowing us to discuss these problems interchangeably.[2] The vast majority of the program follows the same basic logic; to check whether a point z_0 belongs to $\bar{K}_c$, we need to verify whether its trajectory $z_0, p_c(z_0), p_c(p_c(z_0)), \ldots$ escapes to ∞. We pick a (large) number M of iterations. If z_0 does not escape within M steps, we assume that it does not escape. To put this approach in a form consistent with the definition of computability of sets in $\mathbb{R}^2$, let $\mathcal{P}$ be a pixel of size 2^{-n}. The naive decision procedure for determining whether the pixel $\mathcal{P}$ overlaps with $\bar{K}_c$ or is 2^{-n}-far from it thus looks roughly like the following:

A naive heuristic for drawing $\bar{K}_c$:

1. Let z_0 be the center of the pixel $\mathcal{P}$;
2. Let $M = M(n)$ be the number of iterations;
3. **for** $i = 0$ **to** M
    3.1. Set $z_{i+1} \leftarrow z_i^2 + c$;
    3.2. **if** $|z_{i+1}| > |c| + 2$, **return** "$\mathcal{P}$ intersects $\bar{K}_c$";
4. **if** $|z_{M+1}| \leq |c| + 2$ **return** "$\mathcal{P}$ is 2^{-n} far from K_c".

If the pixel $\mathcal{P}$ is evaluated to intersect with $\bar{K}_c$, it is colored black; otherwise it is left white. However, there are multiple problems with these heuristics that make the rendered pictures imprecise and sometimes just wrong. The first is that we take one point z_0 to be "representative" of the entire pixel $\mathcal{P}$. This approach means that even if $\mathcal{P}$ intersects $\bar{K}_c$ or some point $w \in \mathcal{P}$ has its trajectory escape to ∞, we might miss it if the trajectory of z_0 does not escape to ∞. This problem highlights one of the difficulties encountered when developing algorithms for continuous objects. We need to find the answer not just for one point z_0 but for the uncountable set of points located within the pixel $\mathcal{P}$. However, there are computational ways to remedy this problem. Instead of tracing just one point z_0, we can trace the entire geometric shape $\mathcal{P}$, $p_c(\mathcal{P})$, $p_c(p_c(\mathcal{P})), \ldots$ and see whether any part of the Mth iteration of $\mathcal{P}$ escapes to ∞. This step may increase the running time of the algorithm considerably. Nonetheless, we can exploit the peculiarities of complex analytic functions to make the approach work. John Milnor's "Distance Estimator" algorithm does exactly that, at least for a large set of "good" parameters c.

The heuristic also involves a much deeper problem—choosing the parameter $M(n)$. In the thousands of Java applets available online, selection of the number of iterations M is usually left to the user. Suppose we wanted to automate this task or make the program evaluate a "large enough" M such that M iterations are sufficient (see Figure 5 for the effect of selecting an M that is too small); that is, we want to find a parameter M such that if the Mth iteration z_M of z_0 did not escape to ∞, then we can be sure that no further iterations will escape to ∞, and it is safe to assert that z_0 lies within the filled Julia set K_c. Computing such an M is equivalent to establishing termination of the loop

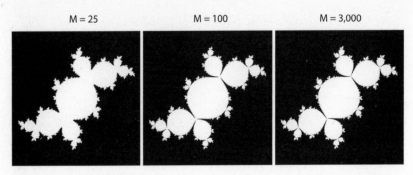

M = 25   M = 100   M = 3,000

FIGURE 5. Example outcomes of the heuristic algorithm with $c \approx -0.126$ + $0.67i$ and $M = 25$, $M = 100$, and $M = 3,000$ iterations. Note that with the lower values of M, the fjords do not reach all the way to the center since the points close to the center do not have time to "escape" in M iterations. The difficulty selecting a "large enough" M is a crucial obstacle in computing the exterior of the filled Julia set K_c.

$$\text{Loop}(z_0): i \leftarrow 0$$
$$\textbf{while } |z_i| \leq |c| + 2$$
$$z_{i+1} \leftarrow z_i^2 + c$$
$$i \leftarrow i + 1$$

In general, the termination of loops, as with the Halting Problem $\mathcal{H}$, is a computationally undecidable problem. If the loop terminates, we are sure that z_0 has escaped. But if the loop keeps running, there is no general way of knowing that it will not terminate later. Thus, there is no simple solution for figuring out the appropriate M; the only way to know that the loop does not terminate is to understand the system X^c and the set $\bar{K}_c$ well enough. Turning the naive heuristic into an algorithm necessarily involves a deep understanding of the underlying dynamical system. As with the systems X^{osc} and x^{comp} discussed earlier, it all boils down to understanding the underlying system. Fortunately, complex dynamicists have developed a rich theory around this system since its introduction around 1917 by the French mathematicians Gaston Julia and Pierre Fatou. This knowledge is enough to give precise answers to most questions concerning the computability of K_c, $\bar{K}_c$, and J_c. One can formalize the naive heuristic discussed here and show that (with slight modifications) it works for the vast majority of values of c.

However, it also turns out that there are noncomputable Julia sets.

THEOREM 1. *There exist parameters c such that the Julia set* J_c *is not computable. Moreover, there are such parameters* $c \in \mathbb{C}$ *that can be produced algorithmically* (Braverman and Yampolsky 2009).

That is, one can produce a parameter c such that drawing the picture of J_c is as computationally hard as solving the Halting Problem. What does such a Julia set look like? All sets in Figure 4 are computable because they were produced algorithmically for inclusion here. Unfortunately, Theorem 1 means that we will most likely never know what these noncomputable Julia sets look like.

The negative result is delicate; in a surprising twist, it does not extend to the filled Julia set K_c.

THEOREM 2. *For all parameters c, the filled Julia set* K_c *is computable* (Braverman and Yampolsky 2009).

Is the Universe a Computer?

We have explored three examples of dynamical systems: The first, the harmonic oscillator X^{osc}, is simple; its behavior can be calculated in closed form, and we can answer pretty much any question about the long-term behavior of the harmonic oscillator computationally. The second, computer system X^{comp}, is at the opposite extreme; predicting it is computationally hard, and it is (relatively) easy to show that it is computationally hard through a reduction to the Halting Problem. The third, complex dynamics, requires an involved approach. For some (in fact, most) parameters c, the long-term behavior of the system X^c is easy in almost any sense imaginable. Showing that there are parameters c for which no amount of computational power suffices to compute the Julia set J_c required a full understanding of the underlying dynamical system developed over almost a century.

Our experience with the computability of Julia sets, as well as the relative success of the field of automated verification at solving undecidable problems in practice (Vardi 2011), indicates that there is likely to be a gap between computability in dynamics in the worst case and in the typical case. This gap means it is possible that although questions surrounding many natural systems (such as the N-body problem and protein assembly) are provably noncomputable in the worst case, a typical case in practice is tractable.

A related, interesting possibility is that noncomputable structures in many systems are too delicate to survive the random noise present in all natural systems. Noise is generally viewed as a "prediction destroying" force, in that making predictions in the presence of noise is computationally more difficult. On the other hand, if we are interested in predicting the statistical distribution of possible future states, then noise may actually make the task easier. It is likely that there are natural systems that (if implemented with no noise) would be computationally impossible to predict, but where the presence of noise makes statistical predictions about the system computationally tractable.

Another lesson from the study of the computational properties of Julia sets is that mapping out which of the Julia sets J_c are and which are not computable requires a nuanced understanding of the underlying dynamical system. It is likely that this is the case with other natural dynamical systems; the prerequisite to understanding their computational properties would be understanding their other properties. Indeed, understanding the role (non)computability and computational universality play in natural dynamical systems probably requires significant advances in both real computation and dynamical systems. The role of computational universality—the ability of natural systems to simulate generic computation—in nature is therefore likely to remain one of the most tantalizing open problems in natural philosophy for some time to come.

Bibliographic Notes

The following references include extensive bibliographies for readers interested in computation over the real numbers. Computability of real numbers was first discussed in Turing's seminal paper (1936), which also started the field of computability. There are two main modern visions on computability over the real numbers: computable analysis and the Blum-Shub-Smale (BSS) framework. My presentation here is fully based on the framework of computable analysis, as presented in depth by Weihrauch (2000). The BSS framework is more closely related to algebraic geometry and presented by Blum et al. (1998). I focused on computable analysis because it appears more appropriate for the study of the computational hardness of natural problems over the real numbers. The results on the computability and complexity of Julia sets was presented by Braverman and Yampolsky (2009). Computational universality of dynamical systems is discussed in several sources, including Moore (1990) and Wolfram (2002), but many basic questions remain open.

Acknowledgments

I would like to thank Ankit Garg, Denis Pankratov, and Omri Weinstein for their multiple comments on earlier versions of this article, and for helping with generating some of the figures. Work on this article has been supported in part by an Alfred P. Sloan Fellowship, National Science Foundation awards CCF-0832797 and CCF-1149888, and a Turing Centenary Fellowship from the John Templeton Foundation.

Notes

1. "Julia" is not a first name in this context but rather the last name of the French mathematician Gaston Julia (1893–1978).
2. As mentioned here, the computability of $(\bar{K}_c)$ and J_c is not equivalent to the computability of the filled Julia set K_c.

References

Blum, L., Cucker, F., Shub, M., and Smale, S. *Complexity and Real Computation.* Springer-Verlag, New York, 1998.

Braverman, M., and Yampolsky, M. *Computability of Julia Sets.* Springer Verlag, Berlin and Heidelberg, 2009.

Milnor, J. *Dynamics in one complex variable.* Princeton University Press, Princeton, NJ, 2006.

Moore, C. "Unpredictability and undecidability in dynamical systems." *Physical Review Letters* 64, 20 (May 1990), 2354–57.

Turing, A.M. "On computable numbers, with an application to the Entscheidungsproblem." *Proceedings of the London Mathematical Society 42,* 2 (Nov. 12, 1936), 230–65.

Vardi, M. "Solving the unsolvable." *Commun. ACM 54,* 7 (July 2011), 5.

Weihrauch, K. *Computable Analysis.* Springer-Verlag, Berlin, 2000.

Wolfram, S. *A New Kind of Science.* Wolfram Media, Champaign, IL, 2002.

Chaos at Fifty

ADILSON E. MOTTER AND DAVID K. CAMPBELL

In classical physics, one is taught that given the initial state of a system, all of its future states can be calculated. In the celebrated words of Pierre Simon Laplace, "An intelligence which could comprehend all the forces by which nature is animated and the respective situation of the beings who compose it—an intelligence sufficiently vast to submit these data to analysis . . . for it, nothing would be uncertain and the future, as the past, would be present to its eyes" (Laplace 1902). Or, put another way, the clockwork universe holds true.

Herein lies the rub: *Exact* knowledge of a real-world initial state is never possible—the adviser can always demand a few more digits of experimental precision from the student, but the result will never be exact. Still, until the 19th century, the tacit assumption had always been that approximate knowledge of the initial state implies approximate knowledge of the final state. Given their success describing the motion of the planets, comets, and stars and the dynamics of countless other systems, physicists had little reason to assume otherwise.

Starting in the nineteenth century, however, and culminating with a 1963 paper by MIT meteorologist Edward Lorenz, a series of developments revealed that the notion of deterministic predictability, although appealingly intuitive, is in practice false for most systems. Small uncertainties in an initial state can indeed become large errors in a final one. Even simple systems for which all forces are known can behave unpredictably. Determinism, surprisingly enough, does not preclude chaos.

A Gallery of Monsters

Chaos theory, as we know it today (Ott 2002), took shape mostly during the last quarter of the twentieth century. But researchers had experienced close encounters with the phenomenon as early as the late

1880s, beginning with Henri Poincaré's studies of the three-body problem in celestial mechanics. Poincaré observed that in such systems "it may happen that small differences in the initial conditions produce very great ones in the final phenomena. . . . Prediction becomes impossible" (Poincaré 1914).

Dynamical systems like the three-body system studied by Poincaré are best described in phase space, in which dimensions correspond to the dynamical variables, such as position and momentum, that allow the system to be described by a set of first-order ordinary differential equations. The prevailing view had long been that, left alone, a conventional classical system will eventually settle toward either a steady state, described by a point in phase space; a periodic state, described by a closed loop; or a quasiperiodic state, which exhibits $n > 1$ incommensurable periodic modes and is described by an n-dimensional torus in phase space.

The three-body trajectories calculated by Poincaré fit into none of those categories. Rather, he observed that "each curve never intersects itself, but must fold upon itself in very complex fashion so as to intersect infinitely often each apex of the grid. One must be struck by the complexity of this shape, which I do not even attempt to illustrate," as paraphrased in English in Mandelbrot (1982, p. 414).

What Poincaré refused to draw is now widely known as a homoclinic tangle, a canonical manifestation of chaos that has fractal geometry. (An image of the tangle can be seen in Figure 4 of Nolte 2010.)

Poincaré's results, independent findings by Jacques Hadamard, and experimental hints of chaos seen by their contemporaries were dismissed by many as pathologies or artifacts of noise or methodological shortcomings and were referred to as a "gallery of monsters" (Mandelbrot 1982). It would take almost another century for chaos theory to gain a lasting foothold.

A Serendipitous Discovery

In all likelihood, Lorenz was unfamiliar with Poincaré's work when he began his foray into meteorology in the mid-1900s (Lorenz 1993, p. 133). With undergraduate and master's degrees in mathematics, Lorenz had served as a meteorologist in World War II before completing his doctoral studies in meteorology at MIT and joining the MIT faculty in 1955.

At the time, most meteorologists predicted weather using linear procedures, which were based on the premise that tomorrow's weather is a well-defined linear combination of features of today's weather. By contrast, an emerging school of dynamic meteorologists believed that weather could be more accurately predicted by simulating the fluid dynamical equations underlying atmospheric flows. Lorenz, who had just purchased his first computer, a Royal McBee LGP-30 with an internal memory of 4096 32-bit words, decided to compare the two approaches by pitting the linear procedures against a simplified 12-variable dynamical model. (Lorenz's computer, though a thousand times faster than his desk calculator, was still a million times slower than a current laptop.)

Lorenz searched for nonperiodic solutions, which he figured would pose the biggest challenge for the linear procedures, and eventually found them by imposing an external heating that varied with latitude and longitude—as does solar heating of the real atmosphere. Sure enough, the linear procedures yielded a far-from-perfect replication of the result.

Having found the nonperiodic solutions of his model interesting in their own right, Lorenz decided to examine them in more detail. He reproduced the data, this time printing the output variables after each day of simulated weather. To save space, he rounded them off to the third decimal place, even though the computer calculations were performed with higher precision. What followed is best appreciated in Lorenz's own words:

> At one point I decided to repeat some of the computations in order to examine what was happening in greater detail. I stopped the computer, typed in a line of numbers that it had printed out a while earlier, and set it running again. I went down the hall for a cup of coffee and returned after about an hour, during which time the computer had simulated about two months of weather. The numbers being printed were nothing like the old ones. I immediately suspected a weak vacuum tube or some other computer trouble, which was not uncommon, but before calling for service I decided to see just where the mistake had occurred, knowing that this could speed up the servicing process. Instead of a sudden break, I found that the new values at first repeated the old ones, but soon afterward differed by one and then several units in the last decimal place, and then began to differ in the next

to the last place and then in the place before that. In fact, the differences more or less steadily doubled in size every four days or so, until all resemblance with the original output disappeared somewhere in the second month. This was enough to tell me what had happened: the numbers that I had typed in were not the exact original numbers, but were the rounded-off values that had appeared in the original printout. The initial round-off errors were the culprits; they were steadily amplifying until they dominated the solution.

(Lorenz 1993, p. 134)

The Butterfly Effect

What Lorenz had observed with his model came to be known as sensitive dependence on initial conditions—a defining property of chaos. In phase space, the phenomenon has a distinct quantitative signature: The distance between any two nearby trajectories grows exponentially with time. Sensitive dependence is illustrated in Figure 1, one of Lorenz's own plots, which shows the gradual divergence of two time series calculated using identical equations but slightly different initial conditions. That trademark behavior gives chaotic systems the appearance of randomness. But as Lorenz himself noted, the appearances are deceiving: At any given time in a random system, one of two or more things can happen next, as one usually assumes for the throw of a die;

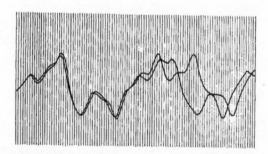

FIGURE 1. The butterfly effect. A close-up of Lorenz's original printout from his discovery of the butterfly effect shows two time series generated with the same equations but with slightly different initial conditions. The series diverge exponentially with time because of sensitive dependence on initial conditions. (Adapted from Gleick 1987.)

FIGURE 2. The Lorenz attractor, as revealed by the never-repeating trajectory of a single chaotic orbit. The spheres shown here represent iterations of the so-called Lorenz equations, calculated using the original parameters in Edward Lorenz's seminal work. From certain angles, the two lobes of the attractor resemble a butterfly, a coincidence that helped earn sensitive dependence on initial conditions its nickname—the butterfly effect. An animated visualization of the attractor is available at http://www.youtube.com/watch?v=iu4RdmBV dps. See also color image. (Image courtesy of Stefan Ganev.)

in chaotic systems such as Lorenz's, outcomes are fully deterministic. (And strictly speaking, so are those of die throws.)

Lorenz realized that if the atmosphere were to behave like his model, forecasting the weather far in the future would be impossible. At a 1972 meeting of the American Association for the Advancement of Science, in a talk titled "Predictability: Does the flap of a butterfly's wings in Brazil set off a tornado in Texas?" Lorenz used a butterfly as a metaphor for a tiny, seemingly inconsequential perturbation that could change the course of weather. The metaphor caught on, and sensitive dependence famously came to be dubbed "the butterfly effect."

Given that computer simulations generally introduce round-off error at each time step—error that is amplified by chaos—one must ask whether Lorenz's solutions can possibly provide reliable information about real chaotic trajectories. As it happens, they can because of a property now known as shadowing: Although for any given initial condition the numerical trajectory diverges from the exact one, there always exists a nearby initial condition whose exact trajectory is approximated by the numerical one for a prespecified stretch of time. In the end, it is as if one had started from a different initial condition and calculated the trajectory exactly—a crucial result, given that numerical calculations came to be widely used in the study of chaotic systems. For example, the trajectories Lorenz calculated using the truncated variables were, in fact, just as representative of his model's behavior as the original (as well as the exact) trajectories.

Lorenz first presented the results from his 12-variable model at a 1960 symposium held in Tokyo. At that meeting, he only briefly mentioned the unexpected effect of round-off errors; he believed that those results belonged in a different paper. In retrospect, he was in little danger of being scooped—apparently, most of his contemporaries failed to recognize the broad significance of his findings. (Meanwhile, the work of other pioneers of chaos often went unappreciated; see Yoshisuke Ueda's description of his frustration at the lack of appreciation of his 1961 analog computer observations of the "randomly transitional phenomenon," later recognized as chaos (Ueda 2001).)

The Lorenz Attractor

Lorenz published his serendipitous discovery in a March 1963 paper titled "Deterministic nonperiodic flow" (Lorenz 1963). He had spent a significant part of his time since the Tokyo meeting looking for the simplest possible model exhibiting sensitive dependence on initial conditions, and he eventually arrived at a three-variable system of nonlinear ordinary differential equations now known as the Lorenz equations (see the box on page 276).

Like Poincaré's three-body system, the Lorenz equations yield phase-space trajectories that never retrace themselves and that don't trace out surfaces of integer dimension. Rather, typical trajectories tend to converge to, and then orbit along, a bounded structure of noninteger, fractal dimension known as a chaotic attractor (Figure 2).

Perhaps the most studied objects in chaos theory, chaotic attractors tend to emerge when a dissipative system is regularly forced to compensate for the loss of energy—as when a child in a swing kicks his or her legs to keep the motion going. In the case of the Lorenz system, forcing is by way of heating, and dissipation is caused by the viscosity of the fluid.

The Lorenz Equations

The three-equation model used by Edward Lorenz to demonstrate chaos derives from a truncated Fourier series expansion of the partial differential equations describing a thin, horizontal layer of fluid heated from below and cooled from above. Lorenz proposed the equations as a crude model of the motion of a region of the atmosphere driven by solar heating of Earth. In standard notation, the equations are $dX/dt = \sigma(-X + Y)$, $dY/dt = rX - Y - XZ$, and $dZ/dt = -bZ + XY$, where X represents the intensity of the convective motion, Y is proportional to the temperature difference between the ascending and descending convective currents, and Z indicates the deviation of the vertical temperature profile from linearity. The parameters b and σ capture particulars of the flow geometry and rheology, and r, the Rayleigh number, determines the relative importance of conductive and convective heat transfer.

Lorenz fixed b and σ at 8/3 and 10, respectively, leaving only r to vary. For small r, the system has a stable fixed point at $X = Y = Z = 0$, corresponding to no convection. At $r = 1$, two symmetrical fixed points, representing two steady convective states, emerge. For $r \geq 24.74$, the convective states lose stability, and at $r = 28$, the system exhibits nonperiodic trajectories like the one shown in Figure 2. Such trajectories forever orbit along a bounded region of the three-dimensional space known as a chaotic attractor and never intersect themselves—otherwise they would be periodic. For larger values of r, the Lorenz equations exhibit a remarkable array of different behaviors, which are carefully cataloged in Sparrow (1982).

A chaotic attractor is the example par excellence of a chaotic set. A chaotic set has uncountably many chaotic trajectories; on such a set, any point that lies in the neighborhood of a given point will also, with probability one, give rise to a chaotic trajectory. Yet no matter the proximity of those two points, in the region between them will lie points of infinitely many periodic orbits. In mathematical parlance, the periodic orbits constitute a countable, zero-measure, but dense set embedded in the chaotic set, analogous to the rational numbers embedded in the set of real numbers. Not only will trajectories that lie on the attractor behave chaotically, any point lying within the attractor's basin of attraction will also give rise to chaotic trajectories that converge to the attractor.

If chaotic sets such as the Lorenz attractor are embedded with infinitely many periodic orbits, why doesn't one ever see those orbits in practice? The answer, and the key feature underlying chaos, is that the periodic orbits are unstable; they cause nearby orbits to diverge, just as the trajectories of a simple pendulum diverge in the neighborhood of the unstable "up" position. But whereas the pendulum trajectories diverge at one point, periodic orbits embedded in the chaotic set cause trajectories to diverge at every point. That skeleton of unstable periodic orbits is what leads to the irregular, chaotic dynamics seen in Lorenz's model and other chaotic systems. Lorenz appears to have grasped that essential feature of chaos early on; he recognized not only that nonperiodicity implies sensitive dependence but also that sensitive dependence is the root cause of nonperiodicity.

One might have expected Lorenz's seminal publication—a model of clarity and concision—to have attracted immediate attention. It did not. Twelve years after its publication, the paper had accumulated fewer than 20 citations. The turning point was when mathematicians and physicists learned of the work, largely through Tien-Yien Li and James Yorke's 1975 paper, "Period three implies chaos" (Li and Yorke 1975), which established the name of the field, albeit with a slightly more restrictive meaning than it has today. By the late 1980s, not only had research on chaos skyrocketed, as evidenced by the thousands of scientific publications on the topic, it was already being widely popularized among nonscientists (Gleick 1987).

Fractals, Folding, and Mille Feuille

Chaotic attractors are generally fractals. The relationship between the chaotic and fractal aspects can be understood by considering the trajectories of a blob of points in the phase space near a chaotic attractor. The chaotic dynamics on the attractor stretches the blob in some directions and contracts it in others, thus forming a thin filament. But because the trajectories are bounded, the filament must eventually fold on itself. When

FIGURE 3. Chaos in dissipative systems. A) The phase-space trajectories of a periodically driven, damped pendulum converge to a chaotic attractor, plotted here at integer multiples of the driving period; stretching and folding of volumes in phase space gives the attractor its fractal structure. B) For a sufficiently dissipative pendulum, the phase space contains two nonchaotic, periodic attractors, indicated here with white dots on the plane of initial conditions. Nevertheless, the phase space contains a chaotic set at the boundary between the attractors' respective basins of attraction. In both panels, the x and y dimensions are position (angle) and angular momentum, respectively. See also color images. (Adapted from Tél and Gruiz 2006 color plate VII. Copyright © 2006 Tamás Tél and Márton Gruiz. Reprinted with permission of Cambridge University Press.)

that sequence of stretching and folding is repeated indefinitely—analogous to a baker kneading dough or preparing *mille feuille* pastry—it gives rise to a fractal set for which the distance between two typical points from the original blob, measured along the resulting attractor, is infinite.

An attractor's geometry can be quantitatively related to its dynamical properties: The (fractal) dimension can be extracted, for example, from the rate at which nearby trajectories diverge in phase space or from the time series of a single variable (Kaplan and Yorke 1979; Packard et al. 1980). Physically, the fractal dimension represents the effective number of degrees of freedom a system has once it has settled on the attractor. Although Lorenz could not resolve it with his numerics, his attractor has a fractal dimension of roughly 2.06.

Figure 3A shows the asymptotic behavior of the phase-space trajectories of another chaotic attractor, that of the periodically driven, damped pendulum. The fractal nature of the attractor can be seen by zooming in on a small portion of phase space: On magnification, the attractor appears statistically self-similar. Given the intimate relationship between the chaotic and fractal natures of such attractors, it was more than a coincidence that the study of fractals reached its maturity in the 1970s, just as chaos was becoming widely known.

Chaos can also find its way into dissipative systems whose attractors are not chaotic, as is the case for a periodically driven pendulum that's very strongly damped. The phase space of such a system, depicted in Figure 3B, contains two periodic attractors, corresponding to clockwise and counterclockwise rotations of the pendulum. Typical trajectories converge to one of the two attractors with probability one; each attractor has its own distinct basin of attraction, as illustrated in the figure. However, embedded at the boundary between those basins of attraction there is a zero-measure, fractal chaotic set—a repeller—that transiently influences the evolution of nearby trajectories. A similar phenomenon may occur in systems whose trajectories are unbounded, as in chaotic scattering processes.

Hamiltonian Chaos

As foreshadowed by Poincaré, chaos can also appear in conservative systems, such as those described by Hamiltonians. Unlike in dissipative systems, where a high-dimensional basin of attraction may converge to a lower dimensional attractor, in conservative systems trajectories necessarily conserve volume in phase space.

To understand how chaos arises in a conservative system, consider a Hamiltonian system—a chain of frictionless harmonic oscillators, say—with n degrees of freedom. The system is integrable, and hence nonchaotic, if it has n independent integrals of motion—that is, if it is described by n conserved quantities such as energy and momenta. If the trajectories are bounded, the system's motion will be constrained to surfaces that are topologically equivalent to n-dimensional tori; each dimension of a torus is associated with a periodic mode of the system. A generic perturbation of the Hamiltonian will destroy resonant tori, for which the various periodic modes have frequency ratios that are easily approximated by rational numbers. Some of the corresponding orbits gain access to $2n$-dimensional regions of the phase space and become chaotic; others form new families of smaller scale tori. The resonant tori in the new families are destroyed by the same mechanism, and so on. The Kolmogorov-Arnold-Moser theorem guarantees that nonresonant tori survive the perturbation, but the fraction of tori, and hence orbits, that fall into that category decreases with the strength of the perturbation. The end result is that the phase space of a generic Hamiltonian system contains coexisting regular and chaotic regions, which extend to arbitrarily small scales (Figure 4).

Beautiful manifestations of Hamiltonian chaos are visible in the asteroid belt and in the rings of Saturn, where unpopulated gaps correspond to chaotic trajectories that were unconfined to the nearly circular, ringlike orbits (Figure 5A).

FIGURE 4. Chaos in conservative systems. A) A stroboscopic map shows the phase-space trajectories of a periodically kicked rotor. The map displays periodic and quasiperiodic regions, which correspond to the looped trajectories in the image, and chaotic regions, which correspond to the scattered trajectories. B) Magnification of the small boxed region in the phase space illustrates the approximately self-similar nature of the phase space. In both panels, the q and p dimensions are position and angular momentum, respectively. See also color images. (Adapted from Campbell 1989.)

A

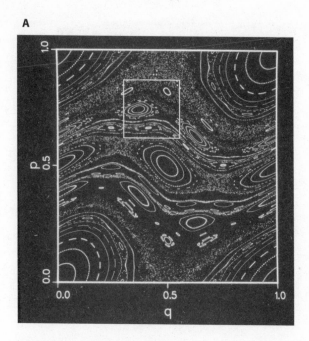

B

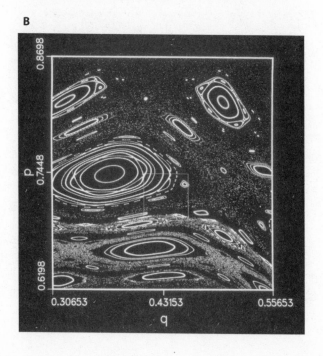

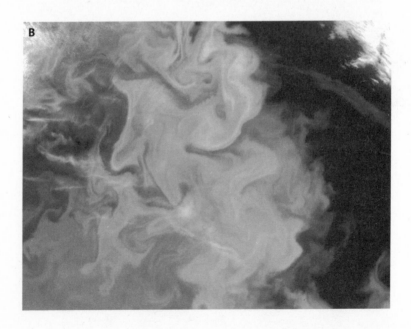

FIGURE 5. Chaos manifests itself in a diverse range of natural settings, including A) the rings of Saturn, where unpopulated gaps correspond to chaotic orbits, as predicted by the Kolmogorov-Arnold-Moser theory (courtesy of NASA/Cassini mission); B) phytoplankton blooms—seen here in a satellite image of the Barents Sea—which form fractallike structures because of chaotic advection (adapted from NASA/Ocean Color Web); and C) Earth's geomagnetic field, which, on astronomical time scales, reverses its poles at irregular, chaotic intervals—a behavior captured by this simulated chaotic attractor (courtesy of Christophe Gissinger). See also color images.

Bifurcations and Universality

Dynamical systems commonly exhibit bifurcations—sudden changes in behavior as a parameter of the system is varied—such as the sudden onset of convection rolls in a fluid heated from below once the temperature gradient exceeds some threshold. A decisive moment in the development of chaos theory came in the late 1970s, when high-precision experimental methods in fluids (Swinney and Gollub 1978) and novel numerical and statistical-physics techniques allowed researchers to explore in quantitative detail how chaos can arise through various sequences of bifurcations.

Mitchell Feigenbaum showed in 1978 that for a wide class of mathematical and experimental systems, one such sequence of bifurcations—the so-called period-doubling route to chaos—occurs the same way, at the same normalized values of the bifurcation parameter. That particular form of universality was subsequently demonstrated in low-temperature convection experiments by Albert Libchaber and Jean Maurer, a development that sparked an explosion of interest in chaos and earned Feigenbaum and Libchaber the 1986 Wolf Prize in Physics (Feigenbaum 1978; Libchaber and Mauer 1980). Since then, theoretical and experimental studies have confirmed the universality of period doubling in a variety of systems, including in the Lorenz equations themselves.

What Have We Learned?

Chaos sets itself apart from other great revolutions in the physical sciences. In contrast to, say, relativity or quantum mechanics, chaos is not a theory of any particular physical phenomenon. Rather, it is a paradigm shift of all science, which provides a collection of concepts and methods to analyze a novel behavior that can arise in a wide range of disciplines. Those traits partly explain the indifference with which the initial hints of the phenomenon were greeted: Early encounters with chaos took place in disparate disciplines—celestial mechanics, mathematics, and engineering—whose practitioners were not aware of each other's findings. Also, chaos generally defies direct analytic approaches. Only when advances in interactive computation made experimental mathematics (Campbell et al. 1985) a reality could one pursue the insights of Poincaré and the other pioneers.

The basics of chaos have been incorporated into physics and applied mathematics curricula, but strong interest remains in understanding specific manifestations of the phenomenon in fields ranging from applied physics and engineering to physiology, computer science, and finance (*Chaos* 2012). For instance, a study reexamining a long-standing debate suggests that a healthy human heartbeat is chaotic because of coupling with breathing (Wessel et al. 2009), much as a star-planet system can become chaotic in the presence of a second planet.

Another body of research has established that, despite sensitive dependence on initial conditions, coupled chaotic systems can synchronize on a shared chaotic trajectory (Pecora and Carroll 1990), a phenomenon with many applications in networked systems (Motter and Albert 2012). Other work has established relations between chaos and the so-called P versus NP problem in computer science. In particular, it has been shown that constraint-optimization problems can be described in terms of dynamical systems that become transiently chaotic as optimization hardness increases (Ercsey-Ravasz and Toroczkai 2011).

Perhaps no field of research can benefit as much from the study of chaos as fluid dynamics. Even in flows governed by periodic velocity fields, microscale fluid elements often move chaotically. A classic example is the transient chaotic behavior of a flow past an obstacle, a behavior that has been proposed to explain how competing plankton species coexist in certain island locales. In a well-mixed environment, all but a handful of species would go extinct. But in the flows that emerge in an island's wake, the various species can inhabit different fractallike flow structures of high surface-to-volume ratio that may intertwine but do not mix (Károlyi et al. 2000) (Figure 5B). Similarly, stretching, folding, and the exponential separation of nearby points—all hallmarks of chaos—are observed in Lagrangian coherent structures, which are of interest, for example, to forecasting contaminant transport in the ocean and atmosphere (Peacock and Haller 2013).

Although low-dimensional chaos does not speak directly to turbulence, spatiotemporal chaos is observed in flows at high Reynolds numbers. Fittingly, Lorenz made the connection between chaos and turbulence at the very beginning—his first choice for the title of his seminal 1963 paper was, in fact, "Deterministic turbulence," which he abandoned only at the urging of the editor.

Numerous fundamental problems in chaos remain at least partially unsettled. They range from the implications of chaos in quantum and relativistic systems to the connection between chaos, coarse graining in phase space, and statistical mechanics. Another fundamental activity concerns model building. For example, the irregular polarity reversals observed at astronomical time scales in Earth's magnetic field have recently been described with a deterministic chaotic model not unlike the three-equation model that begat the field a half-century ago (Figure 5C).

The Lorenz attractor has turned out to be representative of the asymptotic dynamics of many systems, and Lorenz's signature contribution has reverberated both broadly and deeply. As summarized in the citation of his 1991 Kyoto Prize, "He made his boldest scientific achievement in discovering 'deterministic chaos,' a principle which has profoundly influenced a wide range of basic sciences and brought about one of the most dramatic changes in mankind's view of nature since Sir Isaac Newton."

Note

There have been many other important developments in chaos that could not be discussed in this brief, nontechnical article. We offer our apologies to those colleagues whose contributions we were not able to properly acknowledge, and we hope that the article will stimulate readers from other fields to look more deeply into this fascinating subject.

References

D. K. Campbell, *From Cardinals to Chaos: Reflections on the Life and Legacy of Stanislaw Ulam*, N. G. Cooper, ed., Cambridge U. Press, New York (1989).

D. Campbell et al., "Experimental Mathematics: The Role of Computation in Nonlinear Science." *Commun. ACM*, **28**, 374 (1985).

Chaos, special issue, "Fifty Years of Chaos: Applied and Theoretical," *Chaos* **22**(4) (2012).

M. Ercsey-Ravasz and Z. Toroczkai, "Optimization Hardness as Transient Chaos in an Analog Approach to Constraint Satisfaction." *Nat. Phys.*, **7**, 966 (2011).

M. J. Feigenbaum, "Quantitative Universality for a Class of Nonlinear Transformations." *J. Stat. Phys.*, **19**, 25 (1978).

J. Gleick, *Chaos: Making a New Science*, Viking, New York (1987).

J. L. Kaplan and J. A. Yorke, in *Functional Differential Equations and Approximation of Fixed Points,* H.-O. Peitgen, H.-O. Walther, eds., Springer, Berlin (1979), 204.

G. Károlyi et al., "Chaotic Flow: The Physics of Species Coexistence." *Proc. Natl. Acad. Sci, USA,* **97,** 13661 (2000).

P. S. Laplace, *A Philosophical Essay on Probabilities,* F. W. Truscott, F. L. Emory, trans., Wiley, New York (1902), 4. Originally published as *Théorie Analytique des Probabilités,* 6th ed., Madame Veuve Courcier, Paris (1820).

T.-Y. Li and J. A. Yorke, "Period Three Implies Chaos." *Am. Math. Mon.,* **82,** 985 (1975).

A. Libchaber and J. Mauer, "Une Experience de Rayleigh–Benard de Geometrie Reduite; Multiplication, Accrochage et Demultiplication de Frequences." *J. Phys. (Paris) Colloq.,* **41** (C3), 51 (1980).

E. N. Lorenz, "Deterministic Nonperiodic Flow." *J. Atmos. Sci.,* **20,** 130 (1963).

E. N. Lorenz, *The Essence of Chaos,* U. Washington Press, Seattle (1993).

B. B. Mandelbrot, *The Fractal Geometry of Nature,* W. H. Freeman, New York (1982).

A. E. Motter and R. Albert, "Networks in Motion." *Physics Today,* **65**(4), 43 (2012).

D. Nolte, "The Tangled Tale of Phase Space." *Physics Today,* **63**(4), 33 (2010).

E. Ott, *Chaos in Dynamical Systems,* Cambridge U. Press, New York (2002).

N. H. Packard et al., "Geometry from a Time Series." *Phys. Rev. Lett.,* **45,** 712 (1980).

T. Peacock and G. Haller, "Lagrangian Coherent Structures: The Hidden Skeleton of Fluid Flows." *Physics Today,* **66**(2), 41 (2013).

L. M. Pecora and T. L. Carroll, "Synchronization in Chaotic Systems." *Phys. Rev. Lett.,* **64,** 821 (1990).

H. Poincaré, *Science and Method,* F. Maitland, transl., Nelson, New York (1914), p. 68; H. Poincaré, *Les Méthodes Nouvelles de la Mécanique Celeste,* vols. 1–3, Gauthier-Villars, Paris (1899).

C. Sparrow, *The Lorenz Equations: Bifurcations, Chaos, and Strange Attractors,* Springer, New York (1982).

H. Swinney and J. Gollub, "The Transition to Turbulence." *Physics Today,* **31**(8), 41 (1978).

T. Tél and M. Gruiz, *Chaotic Dynamics: An Introduction Based on Classical Mechanics,* K. Kulacsy, trans., Cambridge U. Press, New York, 2006.

Y. Ueda, *The Road to Chaos-II,* Aerial Press, Santa Cruz, CA (2001).

N. Wessel, M. Riedl, and J. Kurths, "Is the Normal Heart Rate 'Chaotic' Due to Respiration?" *Chaos,* **19,** 028508 (2009).

Twenty-Five Analogies for Explaining Statistical Concepts

ROBERTO BEHAR, PERE GRIMA,
AND LLUÍS MARCO-ALMAGRO

Introduction

Well-told analogies, anecdotes, and jokes are resources that teachers use to reinforce the transmission of ideas and concepts. What is more, when used correctly, they bring an informal tone to the class that becomes more enjoyable and sparks student interest.

Recognizing the importance of analogies in the teaching and learning process is not new. Donnelly and McDaniel (1993, 2000) highlighted the possibility of using analogies for teaching in general, and the excellent article of Martin (2003) documents their use in teaching statistics. There are articles with compilations of analogies (Chanter 1983; Brewer 1989) and articles dealing with one specific analogy (Feinberg 1971) about Type I and Type II errors and the judicial process. Gelman and Nolan (2002) proposed many excellent ideas for teaching statistics; Cleary (2005) published a review of that book (with an experiment included, although the deadline for participating has surely passed).

The analogies that we present here are those that we use in our introductory courses. We think they can fit especially well in what Meng (2009) calls *happy courses*: "introductory courses that truly inspire students to *happily* learn statistics as a way of scientific thinking for whatever they do." We use some of the analogies consistently for reinforcing concepts that we are introducing: each of us has his favorites, and those always appear when appropriate. Some others are used or not, depending on how the lecture evolves. Sometimes a group of students have particular difficulties in understanding an idea, and an analogy can be useful to clarify it. Other times we use a suitable analogy as the answer of a question posed by a student.

Our list of analogies includes some well-known examples (like the courtroom analogy for illustrating the idea of hypothesis testing), but most of them are generally unknown; they are either our own creation or we have heard them.

Analogies

STATISTICS IS MUCH MORE THAN PERCENTAGES, SPORTS AVERAGES, AND ELECTION POLLS: ICEBERG

Statistics is often confused with those aspects that are most spoken of, such as those that appear in the media. We can compare this vision of statistics with the vision of an iceberg: you see only a small part. Besides percentages or election polls, statistics plays a fundamental role in many fields of knowledge, such as quality control, the development of new medicines, marketing research, sociological studies, economic indicators, and so on. But these are parts that remain hidden to most people.

THE ROLE OF THE VALUES WE USE (MEAN, STANDARD DEVIATION, . . .) IN THE NUMERICAL SYNTHESIS OF DATA: POLICE SKETCH

How does one describe a face to make a police sketch? The untrained person surely provides information that is vague, confusing, and of little help in drafting the face being described. The police are, however, trained to focus on key elements, and they know the language for describing them. It is similar to the numerical synthesis of data. We choose the measures that best describe the information in the overall data set, and we should be able to understand and correctly interpret its values to form a reliable idea about the information contained within the data.

THE AVERAGE IS INSUFFICIENT FOR DESCRIBING THE DATA: THE HEIGHT OF MARTIANS

If we know that Martians have an average height of 50 inches, are they taller or shorter than us Earthlings? They are not necessarily shorter; it could be that some are only a few inches tall and the majority are more than 80 inches tall.

Using Only the Average Can Lead to Serious Confusion: Jokes About Averages

a. If we are about to cross a river and we are told that the average depth is 3 feet, that does not mean we can relax. It is possible that a large section is 1.5 feet deep, whereas another part is 10 feet deep, where we could drown.

b. If one man eats a chicken and another man none, the average chicken that each man has eaten is one half. When we talk about a ranking of countries using per capita income, we are talking about the half chicken that each person eats, but it remains unclear how many starve.

c. If you enter the kitchen and put your head in the oven and your feet in the refrigerator, your body will be at the ideal average temperature.

d. A statistician goes hunting with two mathematicians. They spot a duck. The first mathematician levels his rifle, fires, and misses to the right. The second mathematician levels his rifle, fires, and misses to the left. The statistician turns to his friends and says "looks to me like we got him, boys."

Properties of the Arithmetic Average: Fulcrum That Balances

When playing with a young child on a teeter-totter, it is not balanced if both the adult and the child sit at the edge of it. The adult weighs a lot more than the child, so the adult must sit closer to the fulcrum to balance the teeter-totter. We can see a dotplot as a teeter-tooter with each point having the same weight. The fulcrum where the dotplot (teeter-totter) is balanced is the average of the data.

Interpreting Standard Deviation in the Context of Normal Distribution: A Basketball Game Between Earthlings and Martians

If the heights of Martians follow a normal distribution with a mean of 55 inches and a standard deviation of 8 inches and they play a game against the Earthling team, who will win? The Earthling basketball players will be around 80 inches tall, which is to say that they are 4 standard deviations from the average (supposing that our heights are normally distributed with mean 68 inches and a standard deviation of 3 inches). If the Martians also choose from among the tallest of their species (those being 4 standard deviations from the average), their players will be around 87 inches tall, and thanks to this height advantage, they will probably beat us (although their average height is less).

THAT A VALUE SHOULD BE CONSIDERED AN ANOMALY DOES NOT DEPEND ONLY ON ITS MAGNITUDE: THE VELOCITY OF VEHICLES ON A HIGHWAY

To study the effect of one type of speed limit sign on the velocity of cars traveling on a section of highway, a hidden radar is installed to measure the speed of each vehicle. The average car travels at about 50 mph. One vehicle passes at 10 mph and another at 90 mph. Although both are at equal extremes (they are at the same distance from the average value), the one that travels at 10 mph should not be considered in the study because it is an agricultural vehicle that cannot go faster and the signal was not intended for it. On the other hand, the car traveling at 90 mph paid no attention to the sign and therefore should not be eliminated because this information is relevant to the study.

WE CAN BE MORE SPECIFIC IN SAYING THAT SOMETHING DEPENDS ON CHANCE: TYPES OF SONGS

People who know nothing about Latin music cannot distinguish among a rumba, a salsa, and a bolero; for them, these are just songs. However, the connoisseur knows how to distinguish the different styles: when told that the next song being played is a bolero, she or he already knows its rhythm, speed, and theme. Similarly, to many people, all random variables look alike. However, a person with knowledge of statistics can assign each variable to a family of variability (normal, binomial, Poisson, etc.) and thus anticipate a lot of its properties.

ALL MODELS ARE THEORETICAL: THERE ARE NO PERFECT SPHERES IN THE UNIVERSE

It appears that the most common geometric form in the universe is the sphere. But how many mathematically perfect spheres are there in the universe? The answer is none. Neither the Earth, nor the Sun, nor a billiard ball is a perfect sphere. So, if there are no true spheres, what good are the formulas for ascertaining the area or volume of a sphere? So it is with statistical models in general and, in particular, with a normal distribution. Although one of the most commonplace examples is height distribution, if we were to have at our disposal the height of every adult on the planet, the histogram profile would not correspond to a Gaussian bell curve, not even if the data were stratified by gender, race, or any other characteristic. But the normal distribution model still provides approximate results that are good enough for practical purposes.

Adding up Random k Values with the Same Probability Distribution Is Not the Same as Multiplying One of Them by k: The Weight of a Dozen Eggs

The weight of a dozen eggs (adding up the values of 12 random variables with the same distribution) presents some variability. The weight of an egg chosen by chance and multiplied by 12 (a random variable value multiplied by 12) presents more variability since the chosen egg may be large and it will be as if all the eggs are large, and if the random egg is small, it will be as if all the eggs are small. These are extreme situations that will not occur if 12 eggs are chosen at random and weighed.

Bias and Precision of an Estimator: Impacts on a Target

The center of a target corresponds to the value of an estimated parameter, and the impacts are the values provided by the estimator. If the estimator is unbiased, all the impacts are around the center, and the lower the variability is, the more grouped the impacts appear. This image also underscores the fact that an unbiased estimator is not always the best option: it could be better if the impacts are not around the center of the target but have very small variability than if they are around the center but are a lot more dispersed.

It Is Not About Demonstrating the Null Hypothesis: Court Trials

In a trial, the null hypothesis is innocence. The objective is not to demonstrate that the accused is innocent but to see if the evidence (the data) contradicts this hypothesis. If there is no evidence, the accused cannot be declared guilty, but this does not mean that innocence has been proven.

The 5% *p*-Value as a Boundary Between the Usual and Unusual Is an Arbitrary Value: The Fingers of the Hand

The 5% *p*-value has been consolidated in many environments as a boundary for whether or not to reject the null hypothesis with its sole merit of being a round number. If each of our hands had six fingers, or four, these would perhaps be the boundary values between the usual and unusual.

Taking All Decisions with the Same Probability of Error Is Not Reasonable: Forgetting an Umbrella or Driving on the Left over a Blind Hill

If you leave the house one morning and shortly afterward find out that the probability of rain is 10%, you may decide not to return home for an umbrella. The probability of an error in this decision is 10%, but nobody would accuse you of being foolish. But if you were driving down an infrequently used road and came upon a blind hill with a pothole in your lane, would you drive on the left to avoid it? Few cars use this road, and there is a low probability that another oncoming car will be in the left lane as you pass, but you most likely would not drive on the left because the consequences of an error would be extremely grave. It is not sensible to unify the probability of error when making decisions. In some cases, 10% is reasonable, but in others, not even 1 in 1,000 is acceptable.

Using Reference Distributions: Screening X-Rays

When a doctor looks at a presumably healthy patient's X-ray to see if there is something abnormal, what he or she does is mentally compare it with the X-ray of a healthy person, taking into account that not all healthy people have exactly the same X-ray. The same is true when comparing a test statistic with its reference distribution in hypothesis testing. The reference distribution is the set of X-rays of healthy people that the doctor has in mind, and the test statistic is the patient's X-ray. If the patient's X-ray looks normal in its reference distribution, the doctor says that everything is fine (he or she cannot think the opposite with the available information). Otherwise, the doctor says that there is something strange, something not normal for a healthy person. There are also cases where the doctor doubts (it could be normal but is not frequent) and asks for additional tests. Luckily, when the reference distribution is a known probability distribution and the test statistic is a value, it is possible to quantify the degree of compatibility through the p-value. For doctors, the quantification of their possible doubts is not so easy.

EFFECT OF SAMPLE SIZE ON THE COMPARISON OF TREATMENTS: MAGNIFICATION OF BINOCULARS

In the distance, we see two animals but we do not know if they are two dogs or a dog and a cat. If we use binoculars with low magnification (low sample size), we are not able to ascertain whether they are both dogs. With an increase in magnification, we can distinguish what they are. If we use a telescope that provides a much greater magnification, we may see in great detail the whiskers of both animals, which are certain to be different, although both are dogs. With enough magnification (sample size), we always see differences even though the animals are the same type. A significant difference (we are sure that the whiskers are different) may be irrelevant to what matters.

THE CONCEPT OF CONFIDENCE LEVEL (1): A PERSON WHO TELLS THE TRUTH 95% OF THE TIME

An exact 95% confidence interval is calculated such that it includes the true value of the estimated parameter 95% of the time. We do not know, however, if the interval we have is one of those that are correct or not. It is like a person who tells the truth 95% of the time, but we do not know whether a particular statement is true.

THE CONCEPT OF CONFIDENCE LEVEL (2): NUMBER OF COMPUTERS THAT ESTIMATE CORRECTLY

Students are asked in a computer lab class to generate random numbers from a normal population. From the sample obtained, each is asked to compute a 95% confidence interval for the mean of the population. We then ask students to raise their hand if the 95% confidence interval they just calculated does not include the true value of the mean. Almost no hands are raised. We discuss with the students the fact that, with a 95% confidence level, the true value is captured in 95% of cases. We repeat the procedure with other levels of confidence and corroborate that when using, for example, a 50% confidence level, only around 50% of students get the mean in the interval.

The Sample Size Versus the Size of the Population: A Spoon for Tasting the Soup

At home, we use a small pot to make soup when we are only two people. To taste whether there is enough salt, we use a teaspoon. Some weekends, there are up to 12 members of the family present, and then we use a pot that is six times larger. Should the spoon also be six times larger? No, the size of the spoon (sample) does not increase proportionally to the size of the pot (population).

The Importance of Sample Representation: Stirring the Pot

Something essential when tasting soup is to stir it well to ensure that the content of the spoon is representative of the content of the pot. If it is not stirred and the sample is removed from the top part, it may be that salt has accumulated in this area and therefore the taste will be more salty, when in fact the whole is lacking in salt. The mistake of not stirring is not corrected simply by using a larger spoon because the error is the same: the problem of a sample not being random is not resolved by making it larger.

Increasing the Sample Size Does Not Always Affect the Usefulness of the Estimate: The Depth of a Lake

If you cannot swim but you have to cross a river, surely you want the river to be as shallow as possible (better 2 feet than 4 feet, just in case . . .). But once you have enough data to assert that the river is too deep to cross (say, about 8 feet), collecting more data to get the exact depth is pointless.

Sample Size and Population Variability: Analyzing a Drop of Blood

It only takes one drop of blood to find out what blood group you belong to because all drops are the same type (no variability). It does not matter if it is from a 7-pound baby or her father, who weighs more than 200 pounds; just a drop is enough to determine the blood group. The greater the variability in the population, the larger the sample size necessary for estimating the value of the parameter.

The Difference Between Correlation and Cause and Effect: The Number of Firefighters and the Damage Caused by Fire

There may be a correlation between two variables without them having a causal relationship. It may be that there is a third, hidden variable that is related to the two, for example, the number of firefighters and fire damage (related variable: size of the fire), the size of shoes worn by children and mathematical skills (child's age), milk consumption and the rate of deaths from cancer (development of the country).

Variable Selection in a Regression Equation: Bringing a Consultant to an Exam

Suppose that you have to take an exam that covers 100 different topics and that you do not know any of them. The rules, however, state that you can bring two classmates as consultants. Suppose also that you know which topics each of your classmates is familiar with. If you could bring only one consultant, it is easy to figure out who you would bring: it would be the one who knows the most topics (the variable most correlated with the response). Let us say this is Paul, who knows 85 topics. With two consultants, you might choose Paul first and, for the second option, it seems reasonable to choose the second most knowledgeable classmate (the second most highly correlated variable), for example, Albert, who knows 75 topics. The problem with this strategy is that it may be that the 75 subjects Albert knows are already included in the 85 that Paul knows and, therefore, Albert does not provide any knowledge beyond that of Paul. A better strategy is to select the second not by considering what he or she knows regarding the entire agenda, but by looking for the person who knows more about the topics that the first does not know (the variable that best explains the residuals of the equation for the variables previously entered). It may even happen that the best pair of consultants are not the most knowledgeable, as there may be two who complement each other perfectly in such a way that one knows 55 topics and the other knows the remaining 45, while the most knowledgeable consultant does not perfectly complement anybody.

RESIDUALS SHOULD NOT CONTAIN INFORMATION: A TRASH BAG

Residuals are what remain after removing all the information from the data. Since they should carry no information, we consider them as "trash." It is necessary to make sure that we do not throw out any trash that has value (information) and that can be exploited to better explain the behavior of the dependent variable.

Final Considerations

An old and renowned professor of statistics once said that instead of preparing the formal presentation of the classes and then, on the fly, improvising anecdotes and analogies, it may be better to prepare the latter and improvise the rest of the class. His argument was based on many anecdotes and stories about his former students who mostly remembered only anecdotes, analogies, similes, or dramatizations and, along with the anecdotes, they retained in their memory the concepts that he had attempted to pass on.

Our experience is that in addition to creating a casual atmosphere in class, analogies are effective in helping the students to understand and remember the ideas we want to convey. We hope that some of those listed here may be useful to our colleagues.

References

Brewer, J. K. 1989. Analogies and Parables in the Teaching of Statistics. *Teaching Statistics*, 11, 21–23.

Chanter, D. O. 1983. Some Anecdotes and Analogies for Illustrating Statistical Ideas. *Teaching Statistics*, 5, 14–16.

Cleary, R. J. 2005. Review of Teaching Statistics: A Bag of Tricks. *The American Statistician*, 59, 275.

Donnelly, C. M., and McDaniel, M. A. 1993. Use of Analogy in Learning Scientific Concepts. *Journal of Experimental Psychology: Learning, Memory and Cognition*, 19, 975–87.

Donnelly, C. M., and McDaniel, M. A. 2000. Analogy with Knowledgeable Learners: When Analogy Confers Benefits and Exacts Costs. *Psychonomic Bulletin and Review*, 7, 537–43.

Feinberg, W. E. 1971. Teaching the Type I and II Errors: The Judicial Process. *The American Statistician*, 25, 30–32.

Gelman, A., and Nolan, D. 2002. *Teaching Statistics: A Bag of Tricks*, Cambridge, MA: Oxford University Press.

Martin, M. A. 2003. "It's Like . . . You Know": The Use of Analogies and Heuristics in Teaching Introductory Statistical Methods. *Journal of Statistics Education* [online], 11, 2. Available at http://www.amstat.org/publications/jse/v11n2/martin.html.

Meng, X. 2009. Desired and Feared—What Do We Do Now and over the Next 50 Years? *The American Statistician*, 63, 202–10.

College Admissions and the Stability of Marriage

David Gale and Lloyd S. Shapley

Introduction

The problem with which we shall be concerned relates to the following typical situation: A college is considering a set of n applicants of which it can admit a quota of only q. Having evaluated their qualifications, the admissions office must decide which ones to admit. The procedure of offering admission only to the q best-qualified applicants is not generally satisfactory, for it cannot be assumed that all who are offered admission will accept. Accordingly, in order for a college to receive q acceptances, it generally has to offer to admit more than q applicants. The problem of determining how many and which ones to admit requires some rather involved guesswork. It may not be known (a) whether a given applicant has also applied elsewhere; if this is known, it may not be known (b) how he or she ranks the colleges to which he or she has applied; even if this is known, it will not be known (c) which of the other colleges will offer to admit him or her. A result of all this uncertainty is that colleges can expect only that the entering class will come reasonably close in numbers to the desired quota and will be reasonably close to the attainable optimum in quality.

The usual admissions procedure presents problems for the applicants as well as the colleges. An applicant who is asked to list in the application all other colleges applied for in order of preference may feel, perhaps not without reason, that by telling a college that it is, say, her or his third choice, she or he will be hurting her or his chances of being admitted.

One elaboration is the introduction of the "waiting list," whereby an applicant can be informed that he or she is not admitted but may be

admitted later if a vacancy occurs. This solution introduces new problems. Suppose that an applicant is accepted by one college and placed
on the waiting list of another that he or she prefers. Should he or she
play it safe by accepting the first or take a chance that the second will
admit him or her later? Is it ethical to accept the first without informing
the second and then withdraw his or her acceptance if the second later
admits him or her ?

We contend that the difficulties here described can be avoided. We
shall describe a procedure for assigning applicants to colleges which
should be satisfactory to both groups, which removes all uncertainties,
and which, assuming that there are enough applicants, assigns to each
college precisely its quota.

The Assignment Criteria

A set of n applicants is to be assigned among m colleges, where q_i is the
quota of the ith college. Each applicant ranks the colleges in the order
of her or his preference, omitting only those colleges that she or he
would never accept under any circumstances. For convenience, we assume that there are no ties; thus, if an applicant is indifferent between
two or more colleges, she or he is nevertheless required to list them in
some order. Each college similarly ranks the students who have applied
to it in order of preference, having first eliminated those applicants
whom it would not admit under any circumstances even if it meant
not filling its quota. From these data, consisting of the quotas of the
colleges and the two sets of orderings, we wish to determine an assignment of applicants to colleges in accordance with some agreed-upon
criterion of fairness.

Stated in this way and looked at superficially, the solution may at
first appear obvious. One merely makes the assignments "in accordance
with" the given preferences. A little reflection shows that complications may arise. An example is the simple case of two colleges, A and B,
and two applicants, α and β, in which α prefers A and β prefers B, but
A prefers β and B prefers α. Here, no assignment can satisfy all preferences. One must decide what to do about this sort of situation. On
the philosophy that the colleges exist for the students rather than the
other way around, it would be fitting to assign α to A and β to B. This
suggests the following admittedly vague principle: other things being

equal, students should receive consideration over colleges. This remark is of little help in itself, but we will return to it later after taking up another more explicit matter.

The key idea in what follows is the assertion that—whatever assignment is finally decided on—it is clearly desirable that the situation described in the following definition should *not* occur:

Definition. An assignment of applicants to colleges will be called **unstable** if there are two applicants α and β who are assigned to colleges A and B, respectively, although β prefers A to B and A prefers β to α.

Suppose that the situation described above did occur. Applicant β could indicate to college A that he would like to transfer to it, and A could respond by admitting β, letting α go to remain within its quota. Both A and β would consider the change an improvement. The original assignment is therefore "unstable" in the sense that it can be upset by a college and applicant acting together in a manner which benefits both.

Our first requirement on an assignment is that it not exhibit instability. This immediately raises the mathematical question: will it always be possible to find such an assignment? An affirmative answer to this question will be given in the next section, and though the proof is not difficult, the result seems not entirely obvious, as some examples will indicate.

Assuming for the moment that stable assignments do exist, we must still decide which among possibly many stable solutions is to be preferred. We now return to the philosophical principle mentioned earlier and give it a precise formulation.

Definition. A stable assignment is called optimal if every applicant is at least as well off under it as under any other stable assignment.

Even granting the existence of stable assignments, it is far from clear that there are optimal assignments. However, one thing that is clear is that the optimal assignment, if it exists, is unique. Indeed, if there were two such assignments, then, at least one applicant (by our "no tie" rule) would be better off under one than under the other; hence one of the assignments would not be optimal after all. Thus the principles of stability and optimality will, when the existence questions are settled, lead us to a unique "best" method of assignment.

Stable Assignments and a Marriage Problem

In trying to settle the question of the existence of stable assignments, we were led to look first at a special case, in which there are the same number of applicants as colleges and all quotas are unity. This situation is, of course, highly unnatural in the context of college admissions, but there is another "story" into which it fits quite readily.

A certain community consists of n men and n women. Each person ranks those of the opposite sex in accordance with his or her preferences for a marriage partner. We seek a satisfactory way of marrying off all members of the community. Imitating our earlier definition, we call a set of marriages *unstable* (and here the suitability of the term is quite clear) if under it there are a man and a woman who are not married to each other but prefer each other to their actual mates.

Question. For any pattern of preferences, is it possible to find a stable set of marriages?

Before giving the answer let us look at some examples.

Example 1. The following is the "ranking matrix" of three men, α, β, and γ, and three women, A, B, and C.

	A	B	C
α	1, 3	2, 2	3, 1
β	3, 1	1, 3	2, 2
γ	2, 2	3, 1	1, 3

The first number of each pair in the matrix gives the ranking of women by the men, the second number is the ranking of the men by the women. Thus, α ranks A first, B second, C third, while A ranks β first, γ second, and α third, etc.

There are six possible sets of marriages; of these, three are stable. One of these is realized by giving each man his first choice, thus α marries A, β marries B, and γ marries C. Note that although each woman gets her last choice, the arrangement is nevertheless stable. Alternatively, one may let the women have their first choices and marry α to C, β to A, and γ to B. The third stable arrangement is to give everyone his or her second choice and have α marry B, β marry C, and γ marry A. The reader will easily verify that all other arrangements are unstable.

Example 2. The ranking matrix is the following.

	A	B	C	D
α	1, 3	2, 3	③, ②	4, 3
β	1, 4	4, 1	3, 3	②, ②
γ	②, ②	1, 4	3, 4	4, 1
δ	4, 1	②, ②	3, 1	1, 4

There is only the one stable set of marriages indicated by the circled entries in the matrix. Note that in this situation no one can get his or her first choice if stability is to be achieved.

Example 3. A problem similar to the marriage problem is the "problem of the roommates." An even number of boys wish to divide up into pairs of roommates. A set of pairings is called *stable* if under it there are no two boys who are not roommates and who prefer each other to their actual roommates. An easy example shows that there can be situations in which there exists no stable pairing. Namely, consider boys α, β, γ, and δ, where α ranks β first, β ranks γ first, γ ranks α first, and α, β, and γ all rank δ last. Then regardless of δ's preferences, there can be no stable pairing, for whoever has to room with δ will want to move out, and one of the other two will be willing to take him in.

The above examples would indicate that the solution to the stability problem is not immediately evident. Nevertheless,

Theorem 1. There always exists a stable set of marriages.

Proof. We shall prove existence by giving an iterative procedure for actually finding a stable set of marriages.

To start, let each boy propose to his favorite girl. Each girl who receives more than one proposal rejects all but her favorite from among those who have proposed to her. However, she does not accept him yet but keeps him on a string to allow for the possibility that someone better may come along later.

We are now ready for the second stage. Those boys who were rejected now propose to their second choices. Each girl receiving proposals chooses her favorite from the group consisting of the new proposers and the boy on her string, if any. She rejects all the rest and again keeps the favorite in suspense.

We proceed in the same manner. Those who are rejected at the second stage propose to their next choices, and the girls again reject all but the best proposal they have had so far.

Eventually (in fact, in at most $n^2 - 2n + 2$ stages) every girl will have received a proposal, for as long as any girl has not been proposed to there will be rejections and new proposals, but since no boy can propose to the same girl more than once, every girl is sure to get a proposal in due time. As soon as the last girl gets her proposal, the "courtship" is declared over, and each girl is now required to accept the boy on her string.

We assert that this set of marriages is stable. Namely, suppose John and Mary are not married to each other but John prefers Mary to his own wife. Then John must have proposed to Mary at some stage and subsequently been rejected in favor of someone that Mary liked better. It is now clear that Mary must prefer her husband to John and there is no instability.

The reader may amuse him or herself by applying the procedure of the proof to solve the problems of Examples 1 and 2, or the following example, which requires ten iterations:

	A	B	C	D
α	1, 3	2, 2	3, 1	4, 3
β	1, 4	2, 3	3, 2	4, 4
γ	3, 1	1, 4	2, 3	4, 2
δ	2, 2	3, 1	1, 4	4, 1

The condition that there be the same number of boys and girls is not essential. If there are b boys and g girls with $b < g$, then the procedure terminates as soon as b girls have been proposed to. If $b > g$, the procedure ends when every boy is either on some girl's string or has been rejected by all of the girls. In either case, the set of marriages that results is stable.

It is clear that there is an entirely symmetrical procedure, with girls proposing to boys, which must also lead to a stable set of marriages. The two solutions are not generally the same as shown by Example 1; indeed, we shall see in a moment that when the boys propose, the result is optimal for the boys, and when the girls propose it is optimal for the girls. The solutions by the two procedures will be the same only when there is a unique stable set of marriages.

Stable Assignments and the Admissions Problem

The extension of our "deferred-acceptance" procedure to the problem of college admissions is straightforward. For convenience, we assume that if a college is not willing to accept a student under any circumstances, as described in the section called "The Assignment Criteria," then that student will not even be permitted to apply to the college. With this understanding, the procedure follows: First, all students apply to the college of their first choice. A college with a quota of q then places on its waiting list the q applicants who rank highest, or all applicants if there are fewer than q, and rejects the rest. Rejected applicants then apply to their second choice, and again each college selects the top q from among the new applicants and those on its waiting list, puts these on its new waiting list, and rejects the rest. The procedure terminates when every applicant is either on a waiting list or has been rejected by every college to which she or he is willing and permitted to apply. At this point, each college admits everyone on its waiting list and the stable assignment has been achieved. The proof that the assignment is stable is entirely analogous to the proof given for the marriage problem and is left to the reader.

Optimality

We now show that the "deferred acceptance" procedure just described yields not only a stable but an optimal assignment of applicants. That is,

Theorem 2. Every applicant is at least as well off under the assignment given by the deferred acceptance procedure as she or he would be under any other stable assignment.

Proof. Let us call a college "possible" for a particular applicant if there is a stable assignment that sends him or her there. The proof is by induction. Assume that up to a given point in the procedure no applicant has yet been turned away from a college that is possible for him or her. At this point, suppose that college A, having received applications from a full quota of better-qualified applicants $\beta_1, \ldots, \beta_q$, rejects applicant α. We must show that A is impossible for α. We know that each β_i prefers college A to all the others, except for those that have previously rejected him or her, and hence (by assumption) are impossible for him or her. Consider a hypothetical assignment that sends α to A and everyone else to colleges that are possible for them. At least one of the β_i will have to go to a less desirable place than A. But this arrangement is unstable, since β_i and A could upset it to the benefit of both. Hence the hypothetical assignment is unstable and A is impossible for α. The conclusion is that our procedure only rejects applicants from colleges which they could not possibly be admitted to in any stable assignment. The resulting assignment is therefore optimal.

Parenthetically, we may remark that even though we no longer have the symmetry of the marriage problem, we can still invert our admissions procedure to obtain the unique "college optimal" assignment. The inverted method bears some resemblance to a fraternity "rush week"; it starts with each college making bids to those applicants it considers most desirable, up to its quota limit, and then the bid-for students reject all but the most attractive offer, and so on.

Concluding Remarks

The reader who has followed us this far has doubtless noticed a certain trend in our discussion. In making the special assumptions needed to analyze our problem mathematically, we necessarily moved further away from the original college admission question, and eventually in discussing the marriage problem, we abandoned reality altogether and entered the world of mathematical make-believe. The practical-minded reader may rightfully ask whether any contribution has been made toward an actual solution of the original problem. Even a rough answer to this question would require going into matters which are nonmathematical, and such discussion would be out of place in a journal of mathematics. It is our opinion, however, that some of the ideas introduced here might usefully be applied to certain phases of the admissions problem.

Finally, we call attention to one additional aspect of the preceding analysis that may be of interest to teachers of mathematics. This is the fact that our result provides a handy counterexample to some of the stereotypes with which nonmathematicians believe mathematics to be concerned.

Most mathematicians at one time or another have probably found themselves in the position of trying to refute the notion that they are people with "a head for figures," or that they "know a lot of formulas." At such times, it may be convenient to have an illustration at hand to show that mathematics need not be concerned with figures, either numerical or geometrical. For this purpose, we recommend the statement and proof of our Theorem 1. The argument is carried out not in mathematical symbols but in ordinary English; there are no obscure or technical terms. Knowledge of calculus is not presupposed. In fact, one hardly needs to know how to count. Yet any mathematician will immediately recognize the argument as mathematical, while people without mathematical training will probably find difficulty in following the argument, though not because of unfamiliarity with the subject matter.

What, then, to raise the old question once more, is mathematics? The answer, it appears, is that any argument that is carried out with sufficient precision is mathematical, and the reason that your friends and ours cannot understand mathematics is not because they have no head for figures but because they are unable to achieve the degree of concentration required to follow a moderately involved sequence of inferences. This observation will hardly be news to those engaged in the teaching of mathematics, but it may not be so readily accepted by people outside of the profession. For them, the foregoing may serve as a useful illustration.

Acknowledgments

The work of the first author was supported in part by the Office of Naval Research under Task NRO47-018.

The Beauty of Bounded Gaps

Jordan Ellenberg

Last May, Yitang "Tom" Zhang, a popular math professor at the University of New Hampshire, stunned the world of pure mathematics when he announced that he had proven the "bounded gaps" conjecture about the distribution of prime numbers—a crucial milestone on the way to the even more elusive twin primes conjecture, and a major achievement in itself.

The stereotype, outmoded though it is, is that new mathematical discoveries emerge from the minds of dewy young geniuses. But Zhang is over 50. What's more, he hasn't published a paper since 2001. Some of the world's most prominent number theorists have been hammering on the bounded gaps problem for decades now, so the sudden resolution of the problem by a seemingly inactive mathematician far from the action at Harvard, Princeton, and Stanford came as a tremendous surprise.

But the fact that the conjecture is *true* was no surprise at all. Mathematicians have a reputation of being no-bullshit hard cases who don't believe a thing until it's locked down and proved. That's not quite true. All of us believed the bounded gaps conjecture before Zhang's big reveal, and we all believe the twin primes conjecture even though it remains unproven. Why?

Let's start with what the conjectures say. The prime numbers are those numbers greater than 1 that aren't multiples of any number smaller than themselves and greater than 1; so 7 is a prime, but 9 is not because it's divisible by 3. The first few primes are: 2, 3, 5, 7, 11, and 13.

Every positive number can be expressed in just one way as a product of prime numbers. For instance, 60 is made up of two 2s, one 3, and one 5. (This is why we don't take 1 to be a prime, though some mathematicians have done so in the past; it breaks the uniqueness, because if 1 counts as prime, 60 could be written as $2 \times 2 \times 3 \times 5$ and $1 \times 2 \times 2 \times 3 \times 5$ and $1 \times 1 \times 2 \times 2 \times 3 \times 5 \ldots$)

The primes are the atoms of number theory, the basic indivisible entities of which all numbers are made. As such, they've been the objects of intense study ever since number theory started. One of the very first theorems in number theory is that of Euclid, which tells us that the primes are infinite in number; we will never run out, no matter how far along the number line we let our minds range.

But mathematicians are greedy types, not inclined to be satisfied with mere assertion of infinitude. After all, there's infinite and then there's *infinite*. There are infinitely many powers of 2, but they're very rare. Among the first 1,000 numbers, there are only 10 powers of 2: 1, 2, 4, 8, 16, 32, 64, 128, 256, and 512.

There are infinitely many even numbers, too, but they're much more common: exactly 500 out of the first 1,000. In fact, it's pretty apparent that out of the first x numbers, just about $(1/2)x$ will be even.

Primes, it turns out, are intermediate—more common than the powers of 2 but rarer than even numbers. Among the first x numbers, about $x/\log(x)$ are prime; this is the prime number theorem, proven at the end of the 19th century by Hadamard and de la Vallée Poussin. This means, in particular, that prime numbers get less and less common as the numbers get bigger, though the decrease is very slow; a random number with 20 digits is half as likely to be prime as a random number with 10 digits.

Naturally, one imagines that the more common a certain type of number, the smaller the gaps between instances of that type of number. If you're looking at an even number, you never have to travel farther than 2 numbers forward to encounter the next even; in fact, the gaps between the even numbers are always *exactly* of size 2. For the powers of 2, it's a different story. The gaps between successive powers of 2 grow exponentially, and there are finitely many gaps of any given size; once you get past 16, for instance, you will never again see two powers of 2 separated by a gap of size 15 or less.

Those two problems are easy, but the question of gaps between consecutive primes is harder. It's so hard that, even after Zhang's breakthrough, it remains a mystery in many respects.

And yet we think we know what to expect, thanks to a remarkably fruitful point of view—we think of primes as *random numbers*.

The reason the fruitfulness of this viewpoint is so remarkable is that the viewpoint is so very, very false. Primes are not random! Nothing about them is arbitrary or subject to chance. Quite the opposite—we take them as immutable features of the universe and carve them on the golden records we shoot out into interstellar space to prove to the extraterrestrials that we're no dopes.

If you start thinking really hard about what "random" *really* means, first you get a little nauseated, and a little after that you find you're doing analytic philosophy. So let's not go down that road.

Instead, take the mathematician's path. The primes are not random, but it turns out that in many ways they *act as if they were*. For example, when you divide a random number by 3, the remainder is 0, 1, or 2, and each case arises equally often. When you divide a big prime number by 3, the quotient can't come out even; otherwise, the so-called prime would be divisible by 3, which would mean it wasn't really a prime at all. But an old theorem of Dirichlet tells us that remainder 1 shows up about equally often as remainder 2, just as is the case for random numbers. So as far as "remainder modulo 3" goes, prime numbers, apart from not being multiples of 3, look random.

What about the gaps between consecutive primes? You might think because prime numbers get rarer and rarer as numbers get bigger, that they also get farther and farther apart. On average, that's indeed the case. But what Yitang Zhang just proved is that there are infinitely many pairs of primes that differ by at most 70,000,000. In other words, that the gap between one prime and the next is bounded by 70,000,000 infinitely often—thus, the "bounded gaps" conjecture.

On first glance, this might seem a miraculous phenomenon. If the primes are tending to be farther and farther apart, what's causing there to be so many pairs that are close together? Is it some kind of prime gravity?

Nothing of the kind. If you strew numbers at random, it's very likely that some pairs will, by chance, land very close together. (Figure 1 is a nice illustration of how this works in the plane; the points are chosen independently and completely randomly, but you see some clumps and clusters all the same.)

It's not hard to compute that, if prime numbers behaved like random numbers, you'd see precisely the behavior that Zhang demonstrated.

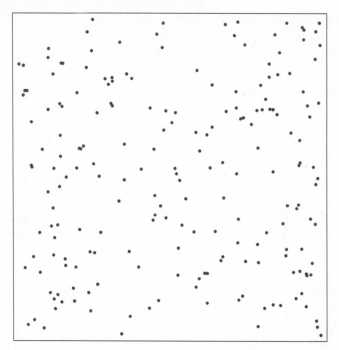

FIGURE 1. A random collection of points exhibits clumps and clusters.

Even more: You'd expect to see infinitely many pairs of primes that are separated by only 2, as the twin primes conjecture claims.

Among the first N numbers, about $N/(\log N)$ of them are primes. If these were distributed randomly, each number n would have a $1/(\log N)$ chance of being prime. The chance that n and $n + 2$ are *both* prime should thus be about $(1/\log N)^2$. So how many pairs of primes separated by 2 should we expect to see? There are about N pairs $(n, n + 2)$ in the range of interest, and each one has a $(1/\log N)^2$ chance of being a twin prime, so one should expect to find about $(N/\log N)^2$ twin primes in the interval.

There are some deviations from pure randomness whose small effects number theorists know how to handle; a more refined analysis taking these into account suggests that the number of twin primes should in fact be about 32 percent greater than $(N/\log N)^2$. This better approximation gives a prediction that the number of twin primes less than a quadrillion should be about 1.1 trillion; the actual figure is 1,177,209,242,304. That's a lot of twin primes.

And a lot of twin primes is exactly what number theorists expect to find no matter how big the numbers get—not because we think there's a deep, miraculous structure hidden in the primes but *precisely because we don't think so.* We expect the primes to be tossed around at random like dirt. If the twin primes conjecture were false, *that* would be a miracle, requiring that some hitherto unknown force would be pushing the primes apart.

Not to pull back the curtain too much, but a lot of famous conjectures in number theory are like this. The Goldbach conjecture that every even number is the sum of two primes? The ABC conjecture, for which Shin Mochizuki controversially claimed a proof last fall? The conjecture that the primes contain arbitrarily long arithmetic progressions, whose resolution by Ben Green and Terry Tao in 2004 helped win Tao a Fields Medal? All are immensely difficult, but they are all exactly what one is guided to believe by the example of random numbers.

It's one thing to know what to expect and quite another to prove that one's expectation is correct. Despite the apparent simplicity of the bounded gaps conjecture, Zhang's proof requires some of the deepest theorems of modern mathematics, like Pierre Deligne's results relating averages of number-theoretic functions with the geometry of high-dimensional spaces. (More classically minded analytic number theorists are already wondering whether Zhang's proof can be modified to avoid such abstruse stuff.)

Building on the work of many predecessors, Zhang is able to show in a precise sense that the prime numbers look random in the first way we mentioned, concerning the remainders obtained after division by many different integers. From this step (following a path laid out by Goldston, Pintz, and Yıldırım, the last people to make any progress on prime gaps), he can show that the prime numbers look random in a totally different sense, having to do with the sizes of the gaps between them. Random is random!

Zhang's success (along with the work of Green and Tao) points to a prospect even more exciting than any individual result about primes— that we might, in the end, be on our way to developing a richer theory of randomness. How wonderfully paradoxical: What helps us break down the final mysteries about prime numbers may be new mathematical ideas that structure the concept of structurelessness itself.

Further Reading

Number theorist Emmanuel Kowalski offers a first report on Zhang's paper: http://bit.ly/10KAyvf.

Here's Terry Tao on the dichotomy between structure and randomness: http://bit.ly/11ZWyd0.

Since Zhang's result appeared in May, several researchers have been working to decrease the size of the bounded gap. Zhang's gap of 70,000,000 is currently down to 5,414. Progress can be tracked here: http://bit.ly/15yECDq.

Contributors

Michael J. Barany is a Ph.D. candidate in the history of science at Princeton University, where he studies the history, sociology, and intellectual and material culture of modern mathematics. A Marshall Scholar and National Science Foundation Graduate Research Fellow, he has published broadly on topics ranging from early modern Euclidean geometry to the sociology of online collaboration in present-day mathematical research. Of particular note, his analyses of the role of stories about counting in Victorian scientific racism have been published by the *British Journal for the History of Science* and the magazine *New Scientist*, and his essay with Donald MacKenzie on the blackboard in mathematical research appears in *Representation in Scientific Practice Revisited* (MIT Press 2014). His dissertation examines the globalization of mathematics and mathematicians in the mid-twentieth century.

Roberto Behar holds a doctoral degree in mathematics and is currently a professor at the Universidad del Valle (in Cali, Colombia). He was one of the founders and first directors of the School of Statistics at this university. He has written a vast number of papers in specialized journals, many of them related to the didactics of statistics, one of his main areas of interest. He has acted as invited speaker and has taught numerous seminars on how to teach mathematics and statistics to undergraduates.

sarah-marie belcastro is a free-range mathematician. Her current positions include director of MathILy, guest faculty at Sarah Lawrence College, and research associate at Smith College. She earned her Ph.D. in mathematics from the University of Michigan and did her undergraduate work in mathematics and astronomy at Haverford College. belcastro's primary mathematical research area is topological graph theory. Among her many non-pure-mathematics interests are the mathematics of knitting, pharmacokinetics, dance (principally ballet and modern), and changing the world. She enjoys connecting people to each other, connecting ideas to each other, and connecting people to ideas. belcastro has written the introductory textbook *Discrete Mathematics with Ducks* and coedited the volumes *Making Mathematics with Needlework: Ten Papers and Ten Projects* and *Crafting by Concepts: Fiber Arts* *and Mathematics.* More information on all of the above and many resources are available at sarah-marie's website http://www.toroidalsnark.net.

Mark Braverman is an assistant professor of computer science at Princeton University, where he has been on the faculty since 2011. He received his Ph.D. from the University of Toronto in 2008 under the supervision of Stephen Cook. In his Ph.D. studies, he worked on computability in complex dynamics, developing the theory of computation for Julia sets. More recently, he has been working on new ways of merging ideas from information theory and computational complexity theory. He is a recipient of a number of awards, including the 2009 Canadian Mathematical Society Doctoral Prize and a 2013 Packard Fellowship in Science and Engineering.

Lawrence Brenton is a professor of mathematics at Wayne State University in Detroit. His current research interests include the geometry of complex manifolds, applications of singularity theory to cosmology, and topics in the history of mathematics and in mathematics education. Professor Brenton is the recipient of the Mathematical Association of America's George Pólya Award (2009) and the Trevor Evans Award (2011) for expository writing.

David K. Campbell is a professor of physics and electrical and computer engineering at Boston University. He is known for his theoretical studies of localized nonlinear excitations—solitons, polarons, breathers, intrinsically localized modes—in physical systems and of strongly correlated electronic materials, including conducting polymers, magnetic systems, high-temperature superconductors, graphene, and related novel solid state systems. An international leader in the field of "nonlinear science," in 2010 he received the American Physical Society's J. Edgar Lilienfeld Prize for his contributions to this field. He is the founding editor-in-chief of the American Institute of Physics' journal *Chaos* and a Fellow of the American Physical Society and the American Association for the Advancement of Science.

John H. Conway is John von Neumann Emeritus Professor of Pure and Applied Mathematics at Princeton University and a Fellow of the Royal Society of London. His contributions span several mathematical fields, including number theory, knot theory, combinatorial game theory, the theory of finite groups, and recreational mathematics. Among the books Conway wrote or cowrote are *Sphere Packings, Lattices and Groups, The Sensual (Quadratic) Form, On Numbers and Games, The Book of Numbers, The Symmetries of Things*, and *Winning Ways for Your Mathematical Plays*. John Conway was awarded the Pólya Prize, the Nemmers Prize in Mathematics, and the Leroy P. Steele Prize of the American Mathematical Society.

Kenneth Cukier is the data editor of *The Economist* in London and was previously its technology correspondent. He is the coauthor with Viktor Mayer-Schönberger of *Big Data: A Revolution That Will Transform How We Live, Work and Think,* published in 2013, which was a *New York Times* bestseller and was translated into 20 languages. His writings have appeared in the *New York Times,* the *Wall Street Journal,* and other newspapers, and he is a frequent commentator on BBC, CNN, and NPR. He also serves on the World Economic Forum's advisory council on data-driven development.

Keith Devlin is a cofounder and executive director of Stanford University's H-STAR Institute and a cofounder of the Stanford Media X research network. He is a World Economic Forum Fellow, a Fellow of the American Association for the Advancement of Science, and a Fellow of the American Mathematical Society. His current research is focused on the use of different media to teach and communicate mathematics to diverse audiences. In this connection, he is a cofounder and president of an educational technology company, BrainQuake, that creates mathematics learning video games. He also works on the design of information and reasoning systems for intelligence analysis. Other research interests include theory of information, models of reasoning, applications of mathematical techniques in the study of communication, and mathematical cognition. He has written 32 books and more than 80 published research articles. He is the recipient of the Pythagoras Prize, the Peano Prize, the Carl Sagan Award, and the Joint Policy Board for Mathematics Communications Award. In 2003, he was recognized by the California State Assembly for his "innovative work and longtime service in the field of mathematics and its relation to logic and linguistics." He is "the Math Guy" on National Public Radio.

Penelope Dunham is a professor of mathematics at Muhlenberg College. Her research and publications focus on technology use in mathematics instruction, including issues of equity, access, and pedagogy. She is a past editor of the Connecting Research to Teaching department of *Mathematics Teacher* and has given workshops on calculator use to university, secondary, and elementary teachers throughout the United States, as well as in Great Britain and Japan. She has twice been honored with awards for distinguished teaching (at Hanover College and Muhlenberg College) and frequently gives presentations on teaching undergraduate mathematics. She holds a master's degree in mathematics and a doctorate in mathematics education from The Ohio State University.

Jordan Ellenberg is Vilas Distinguished Achievement Professor of Mathematics at the University of Wisconsin–Madison. His research centers on the interactions between number theory, algebraic geometry, and algebraic topology. He writes frequently on mathematical topics for newspapers and magazines, including *Slate*, the *New York Times*, and the *Wall Street Journal*. His book *How Not To Be Wrong*, from Penguin Press, is a tour through the ubiquitous appearances of mathematical ideas in everyday life.

David Gale was a professor emeritus at the University of California–Berkeley, with work in mathematical economics, game theory, and convex analysis. In economics, he contributed an early proof of the existence of competitive equilibrium, a solution to the n-dimensional Ramsey problem, and results in the theory of optimal economic growth; in mathematics, he defined an involution on sets of points in projective sets important in optimization, coding theory, and algebraic geometry; he also invented and popularized mathematical games. Among his books are *The Theory of Linear Economic Models* and *Tracking the Automatic Ant*. He was a member of the National Academy of Sciences, the American Academy of Arts and Sciences, and of other prestigious societies. Late in life, Gale developed the award-winning MathSite (http://mathsite.math.berkeley.edu/main.html).

Marshall Gordon received his doctorate from Columbia University Teachers College. He taught at SUNY–Stony Brook and UNC–Greensboro before going to the Park School of Baltimore, where he and colleagues have been writing mathematics curricula with a habits-of-mind orientation that provides valuable perspective for learning and teaching mathematics. His "Mathematical Habits of Mind: Promoting Students' Thoughtful Considerations", *Journal of Curriculum Studies*, 2011, offers a pedagogical rationale and mathematics examples. He is interested in applying John Dewey's philosophy of education to the classroom; his presentation, "Dewey's Criteria of Interaction and Continuity to Evaluate Project-Based Learning" at the John Dewey Society annual meeting, 2014, provides an instance of how valuable Dewey's thinking is to educational practice.

Pere Grima holds a doctoral degree in industrial engineering and is a professor of statistics at Universitat Politecnica de Catalunya, BarcelonaTech (Spain). His area of expertise is statistical methods for quality control and improvement. He has worked as a statistical consultant in private companies and public administration. He is the author of numerous papers and coauthor of the book *Industrial Statistics with Minitab*. He is also interested in the dissemination of statistical thinking, focusing not only on undergraduates but also on high school students and their mathematics teachers. About this last topic, he has written the book *Absolute Certainty and Other Fiction*, in the Everything Is Mathematical Series of books.

Brian Hayes is a senior writer for *American Scientist* magazine and a former editor of both *American Scientist* and *Scientific American*. He is also the author of *Infrastructure: A Field Guide to the Industrial Landscape* (W. W. Norton 2005) and *Group Theory in the Bedroom and Other Mathematical Diversions* (Hill and Wang 2008). His website http://bit-player.org focuses on computational and mathematical topics. He has been journalist-in-residence at the Mathematical Sciences Research Institute in Berkeley, California, and a visiting scientist at the Abdus Salam International Centre for Theoretical Physics in Trieste, Italy. He is the winner of a National Magazine Award.

Bahman Kalantari is a professor of computer science at Rutgers University. His main research areas are theory and algorithms for various optimization problems and for solving polynomial equations. He has coined the term *polynomiography* for algorithmic visualization of polynomial equations and holds a U.S. patent for its technology. Polynomiography has received attention in national and international media. Kalantari believes that if fully developed, polynomiography will democratize polynomials, turning these abstractions into entities that will be appreciated by the general public, will lead to novel applications, and will attract youth to math and science. Over the years, he has taken a keen interest in fine art, even fashion, all based on polynomiography. He is the author of the book *Polynomial Root-Finding and Polynomiography*. He maintains the website www.polynomiography.com.

Tanya Khovanova is a lecturer at MIT and a freelance mathematician. She received her Ph.D. in mathematics from Moscow State University in 1988. At that time, her research interests were in representation theory, integrable systems, super-string theory, and quantum groups. Her research was interrupted by a period of employment in industry, where she became interested in algorithms, complexity theory, cryptography, and networks. Several years ago, she resigned from industry to return to research. Her current interests lie in recreational mathematics, including puzzles, magic tricks, combinatorics, number theory, geometry, and probability theory. Her website is located at tanya khovanova.com, her highly popular math blog at blog.tanyakhovanova.com, and her Number Gossip website at numbergossip.com. Khovanova works with gifted children in a variety of settings. At the Advanced Math & Science Academy Charter School in Marlborough, Massachusetts, she coaches the math competition team. She also mentors and supervises high school students who want to do research in mathematics. She works with the two most successful programs that focus on helping kids do research: RSI and PRIMES-MIT.

Nicole Lazar earned her Ph.D. in statistics from the University of Chicago. After serving on the faculty of the Department of Statistics at Carnegie

Mellon University, she moved to the University of Georgia, where she is currently a professor of statistics. Her main areas of research are likelihood theory, multiple testing problems, and the analysis of functional neuroimaging data. She is the author of *The Statistical Analysis of Functional MRI Data* (2008). Lazar is an elected member of the International Statistical Institute and a Fellow of the American Statistical Association.

Uri Leron is Churchill Family Professor (Emeritus) of Science and Technology Education at the Technion–Israel Institute of Technology. He received his Ph.D. in mathematics (ring theory) from the Hebrew University in Jerusalem and started off his professional career as a research mathematician at the University of Oregon, UCLA, and the Israel Institute of Technology. Halfway through his career, he switched to working in mathematics and computer science education. His professional interests—all aimed at trying to bridge the gap between intuitive and analytical thinking—have included the psychology of mathematical thinking, new formats for mathematical proofs, constructionist uses of digital interactive environments, and the teaching of mathematics and computer science in high schools. Recently he has been interested in implications of mathematical thinking on evolutionary psychology and the "rationality debate" in cognitive psychology. He has published two books and numerous research papers but considers his most important contribution the 20-some graduate students for M.Sc. and Ph.D. degrees he has supervised.

Lluís Marco-Almagro obtained his Ph.D. in industrial statistics in 2011 and is currently an associate professor in the School of Engineering at Universitat Politecnica de Catalunya, BarcelonaTech (Spain). His areas of interest include industrial statistics, especially design of experiments and statistical process control, and statistical methods in effective design of products. He is coauthor of the book *Industrial Statistics with Minitab*. He led several teaching innovation projects with the aim of improving statistical learning in undergraduates.

Viktor Mayer-Schönberger is Professor of Internet Governance and Regulation at the University of Oxford. Before that, he spent a decade on the faculty at Harvard University. He researches the economic, political, and social implications of information networks. He has published about a hundred scientific papers, including ten books, such as the award-winning *Delete: The Virtue of Forgetting in the Digital Age* and (together with Kenneth Cukier) the bestselling *Big Data*, available in twenty languages. Before joining academia, he was a software entrepreneur. He holds degrees (among others) from Harvard and the London School of Economics, sits on boards, and advises international organizations, corporations, and governments on the information economy.

Adilson E. Motter is the Harold H. and Virginia Anderson Professor of Physics and Astronomy at Northwestern University. He is an expert in nonlinear dynamics and network modeling of complex systems, with contributions to the foundations of chaos, network synchronization, cascading failures, network control, and the discovery of synthetic rescues. He is Fellow of the Alfred P. Sloan Research Foundation and of the American Physical Society, and recipient of a National Science Foundation CAREER Award and of the 2013 Erdős-Rényi Prize in Network Science.

Stephen Pollard is a professor of philosophy at Truman State University. He received his B.A. from Haverford College in 1979 and his Ph.D. from the University of Texas in 1983. His research and publications deal primarily with logic and the philosophy of mathematics, but his interests also include classical Greek philosophy, American pragmatism, and the philosophy of science. His books include *Philosophical Introduction to Set Theory* and (with N. M. Martin) *Closure Spaces and Logic*.

Carlo H. Séquin has been on the faculty in the EECS Department at UC Berkeley since 1977. He has been in love with geometry since high school and has typically focused on the geometry component in most of his professional assignments. Those tasks included the development of the first TV-compatible solid-state image sensor at Bell Labs, the "RISC" microcomputer built at UC Berkeley with Dave Patterson and their students, the design effort for Soda Hall—the current home of the Computer Science Division—and the prototyping of computer-aided design (CAD) tools for circuit designers, architects, mechanical engineers, and artists creating abstract geometric art. More recently, Séquin has focused on making visualization models for geometrical and topological concepts, such as regular maps on low-genus surfaces, or regular homotopies of tori and Klein bottles. These models are typically realized with computer-driven, layered-manufacturing machines and often result in aesthetically appealing sculptural maquettes.

Lloyd S. Shapley is a professor emeritus at the University of California—Los Angeles. His work has had a major impact on the development of mathematical economics, especially game theory. Among other contributions, he introduced the Shapley value in the analysis of stochastic games, the Shapley-Shubik power index for weighted or block voting power, and the Bondareva-Shapley theorem implying that convex games have nonempty cores. In 2012, Shapley was awarded the Nobel Prize for Economics.

Joseph Shipman, is a mathematical consultant who lives and works in the Princeton area. He has degrees in mathematics from MIT and Brandeis. In addition to consulting work involving operations research, finance, software development, election polling, and gaming, he teaches and tutors high school mathematics and collaborates on mathematical research with John Conway and others. In 1990, he proved that it is consistent that iterated integrals of nonnegative functions taken in different orders of integration never disagree, even for nonmeasurable functions (along with various necessary conditions and sufficient conditions for the agreement of iterated integrals). In 2007, he proved that any field in which all polynomials of prime degree have roots is algebraically closed (along with a necessary and sufficient condition for all implications between statements of the form "all polynomials of degree n have roots").

Francis Edward Su is the Benediktsson-Karwa Professor of Mathematics at Harvey Mudd College and earned his Ph.D. from Harvard University. He is president-elect of the Mathematical Association of America. His research is in geometric combinatorics and applications to the social sciences, and he has coauthored numerous papers with undergraduates. He also has a passion for teaching and popularizing mathematics. From the Mathematical Association of America, he received the 2001 Hasse Prize for expository writing, and the 2004 Alder Award and the 2013 Haimo Award for distinguished teaching. He authors the popular Math Fun Facts website and iPhone app. His hobbies include songwriting, gardening, photography, and theology. Just like mathematics, these are modes of creative expression that divinely blend structure and freedom, truth and beauty, reflection and action.

John Z. Sun received his Ph.D. in 2013 from the Department of Electrical Engineering & Computer Science at the Massachusetts Institute of Technology. His thesis explored quantization theory and its applications in media compression, communication systems, and neuroscience. He received the best student paper award at the IEEE Data Compression Conference, the Claude E. Shannon Research Assistantship, and the inaugural EECS Paul L. Penfield Student Service Award. He is currently a quantitative researcher at a hedge fund in New York City.

Bruce Torrence is the Garnett Professor of Mathematics at Randolph-Macon College, where he serves as chair of the Department of Mathematics.

In 2008, he received the John Smith Award for distinguished teaching from the Mathematical Association of America. He was a coeditor of *Math Horizons* from 2008 to 2013, and he was cochair of Mathematics Awareness Month in 2014. Trained as an algebraic topologist, he has a long-standing interest in computer algebra systems. His current research focuses on combinatorial games and graph theory.

Lav R. Varshney is an assistant professor in the Department of Electrical and Computer Engineering at the University of Illinois at Urbana–Champaign. He received his B.S. degree from Cornell University in 2004 and his S.M., E.E., and Ph.D. degrees from the Massachusetts Institute of Technology in 2006, 2008, and 2010, respectively. His master's thesis received the E. A. Guillemin Thesis Award, and his doctoral thesis received the J.-A. Kong Award Honorable Mention. He was previously a research staff member at IBM Research, where his work on computational creativity, crowdsourcing, and data analytics was recognized. His current research focus, on the science and engineering of informational systems that involve humans and machines, aims to improve individual and collective intelligence in modern environments.

Orit Zaslavsky has been a professor of mathematics education at New York University since 2009. She received her Ph.D. from the Technion–Israel Institute of Technology in 1987, where she returned as a faculty member after a two-year postdoctorate fellowship at the University of Pittsburgh's Learning Research and Development Center. Her recent work focuses on pupils' and mathematics teachers' understandings of the interplay between mathematical examples, definitions, and proof and how these understandings come to play when involved in mathematical activities. She was a pioneer in studying the development of mathematics teacher-educators, from both theoretical and practical perspectives and investigated and conceptualized the nature of productive mathematics-related tasks and instructional examples. She published numerous journal articles and coedited a Springer book, *Constructing Knowledge for Teaching Secondary Mathematics*, and two special issues on these topics. In addition, she directed major large-scale curriculum design and professional development projects on mathematics education in Israel.

Dov Zazkis received his Ph.D. from the joint doctoral program in mathematics and science education at San Diego State University and the University of California–San Diego. He is currently a postdoctoral researcher at Rutgers University and will be moving to Oklahoma State University in fall 2014. His primary research interest is undergraduate mathematics students' use of semantic (visual) and syntactic (analytic) representations in problem solving and proof.

Rina Zazkis is a professor of mathematics education at Simon Fraser University, in Vancouver, Canada. Her research focuses on the mathematical knowledge of undergraduate students and the ways in which this knowledge is acquired and modified. Teaching, learning, and understanding of abstract algebra, along with elementary number theory are specific focuses of her work. Elementary number theory is summarized in her book *Relearning Mathematics: A Challenge for Prospective Elementary School Teachers*. This research has been continuously funded by the Social Sciences and Humanities Research Council of Canada. *Teaching Mathematics as Storytelling* (2009) and *Lesson Play in Mathematics Education: A Tool for Research and Professional Development* (2013) are her recent coauthored books that introduce innovative ideas in teacher education.

Notable Writings

The following is a bibliography of some of the most interesting 2013 articles on mathematics. At some point during the preparation of this volume, I considered for selection most of the entries on this list.

Acotto, Edoardo, and Moreno Andreatta. "Between Mind and Mathematics: Different Kinds of Computational Representations of Music." *Mathematics and Social Sciences* 50.199(2013): 7–25.

Adams, Colin, Nöel MacNaughton, and Charmaine Sia. "From Doodles to Diagrams to Knots." *Mathematics Magazine* 86(2013): 83–96.

Artigue, Michèle, and Morten Blomhøj. "Conceptualizing Inquiry-Based Education in Mathematics." *ZDM Mathematics Education* 45(2013): 797–810.

Azzouni, Jodi. "The Relationship of Derivations in Artificial Languages to Ordinary Rigorous Mathematical Proof." *Philosophia Mathematica* 21(2013): 247–54.

Baker, Mark, et al. "Sets, Planets, and Comets." *The College Mathematics Journal* 44.4(2013): 258–64.

Barras, Colin. "Elements of Thought." *New Scientist* February 9, 2013.

Barrett, Jeffrey A. "On the Coevolution of Basic Arithmetic Language and Knowledge." *Erkenntnis* 78(2013): 1025–36.

Behrisch, Lars. "Political Economy and Statistics in the *Ancien Régime*." In *Writing Political History Today*, edited by Willibald Steinmetz, Ingrid Gilcher-Holtey, and Heinz-Gerhard Haupt. Frankfurt-am-Main, Germany: Campus Verlag, 2013.

Bennett, Curtis, and Jacqueline Dewar. "SOLT and Interdisciplinary Encounters in the Study of Students' Understanding of Mathematical Proof." In *The Scholarship of Teaching and Learning In and Across the Disciplines*, edited by Kathleen McKinney. Bloomington: Indiana University Press, 2013.

Berman, Bob. "Speaking the Language of the Cosmos." *Astronomy* December 2013, 24–29.

Biggs, Norman. "Thomas Harriot on Continuous Compounding." *British Society for the History of Mathematics Bulletin* 28.2(2013): 66–74.

Blair, Kristen P., Jessica M. Tsang, and Daniel L. Schwartz. "The Bundling Hypothesis: How Perception and Culture Give Rise to Abstract Mathematical Concepts in Individuals." In *International Handbook of Research on Conceptual Change*, edited by Stella Vasniadou, pp. 322–40. New York: Routledge, 2013.

Błaszczyk, Piotr, Mikhail G. Katz, and David Sherry. "Ten Misconceptions from the History of Analysis." *Foundations of Science* 18(2013): 43–74.

Brody, Linda E. "The Promise of Mathematical Precocity." In *The Complexity of Greatness: Beyond Talent and Practice,* edited by Scott Barry Kaufmann. New York: Oxford University Press, 2013.

Brown, Ethan, and Nick Bearman. "Listening to Uncertainty." *Significance* 10(2012): 14–17.

Brown, Tony, et al. "Experiencing the Space We Share: Rethinking Subjectivity and Objectivity." *ZDM Mathematics Education* 45(2013): 561–72.

Brown, Tony, Elaine Hodson, and Kim Smith. "TIMSS Mathematics Has Changed Real Mathematics Forever." *For the Learning of Mathematics* 33.2(2013): 38–43.

Burn, Bob. "Root 2: The Early Evidence and Later Conjectures." *British Society for the History of Mathematics Bulletin* 28(2013): 114–20.

Busch, Jacob. "The Indispensability Argument for Mathematical Realism and Scientific Realism." *Journal for General Philosophy of Science* 43(2012): 3–9.

Carter, Jessica. "Handling Mathematical Objects." *Synthese* 190(2013): 3983–99.

Cellucci, Carlo. "Philosophy of Mathematics: Making a Fresh Start." *Studies in History and Philosophy of Science* 44(2013): 32–42.

Cellucci, Carlo. "Top-Down and Bottom-Up Philosophy of Science." *Foundations of Science* 18(2013): 93–106.

Cole, Julian. "Towards an Institutional Account of the Objectivity, Necessity, and Atemporality of Mathematics." *Philosophia Mathematica* 21(2013): 9–36.

Corry, Leo. "Geometry and Arithmetic in the Medieval Traditions of Euclid's *Elements*." *Archive for History of Exact Sciences* 67(2013): 637–705.

De Cruz, Helen, and Johan De Smedt. "Mathematical Symbols as Epistemic Action." *Synthese* 190(2013): 3–19.

de Freitas, Elizabeth, and Nathalie Sinclair. "New Materialist Ontologies in Mathematics Education: The Body in/of Mathematics." *Educational Studies in Mathematics* 83(2013): 453–70.

de Sojo, Aurea Anguera, er al. "Turing and the Serendipitous Discovery of the Modern Computer." *Foundations of Science* 18(2013): 545–57.

Denning, Peter J. "Thumb Numbers." *Communications of the ACM* 56.6(2013): 35–37.

Devlin, Keith. "Will We Give Up Our Constitutional Freedoms Because We Can't Count?" *Huffington Post* June 11, 2013.

Dieveney, Patrick. "Anything and Everything." *Erkenntnis* 78(2013): 119–40.

Domski, Mary. "Kant and Newton on the A Priori Necessity of Geometry." *Studies in History and Philosophy of Science* 44(2013): 438–47.

Dubinsky, Ed, and Robin T. Wilson. "High School Students' Understanding of the Function Concept." *The Journal of Mathematical Behavior* 32(2013): 83–101.

Eddy, Matthew Daniel. "The Shape of Knowledge: Children and the Visual Culture of Literacy and Numeracy." *Science in Context* 26(2013): 215–45.

Efron, Bradley. "Bayes' Theorem in the 20th Century." *Science* 340(June 7, 2013): 1177–78.

Everson, Michelle, Ellen Gundlach, and Jacqueline Miller. "Social Media and the Introductory Statistics Course." *Computers in Human Behavior* 29(2013): A69–A81.

François, Karen, Kathleen Coessens, and Jean Paul Van Bendegem. "The Space of Mathematics: Dynamic Encounters between Local and Universal." In *Educational Research: The Importance and Effects of Institutional Spaces*, edited by Paul Smeyers, Mark Depaepe, and Edwin Keiner. Dordrecht, Germany: Springer, 2013.

Frank, Martin. "Mathematics, Technics, and Courtly Life in Late Renaissance Urbino." *Archive for History of Exact Sciences* 67(2013): 305–30.

Gabay, Shai, et al. "Size before Numbers: Conceptual Size Primes Numerical Value." *Cognition* 129(2013): 18–23.

Garelick, Barry. "Problem Solving: Moving from Routine to Nonroutine and Beyond." *Notices of the AMS* 60.10(2013): 1340–42.

Gazit, Avikam. "Pre-service Mathematics Teachers' Attitudes toward Integrating Humor in Math Lessons." *The Israeli Journal of Humor Research* 3(2013): 27–44.

Gelman, Andrew, and Mark Palko. "The War on Data." *Chance* 26.1(2013): 57–60.

Gelman, Andrew, and Christian P. Robert. "'Not Only Defended But Also Applied': The Perceived Absurdity of Bayesian Inference." *The American Statistician* 67.1(2013): 1–5.

Giovanelli, Marco. "The Forgotten Tradition: How the Logical Empiricists Missed the Philosophical Significance of the Work of Riemann, Christoffel, and Ricci." *Erkenntnis* 78(2013): 1219–57.

Glaz, Sarah. "Ode to Prime Numbers." *American Scientist* 101.4(2013): 246–50.

Goldstine, Susan, Sophie Sommer, and Ellie Baker. "Beading the Seven-Color Theorem." *Math Horizons* 21.1(2013): 22–24.

Grattan-Guinness, I. "The Mentor of Alan Turing; Max Newman (1897–1984) as a Logician." *The Mathematical Intelligencer* 35.3(2013): 54–63.

Grosholz, Emily R. "Teaching the Complex Numbers: What History and Philosophy of Mathematics Suggest." *Journal of Humanistic Mathematics* 3.1(2013).

Grosslight, Justin. "Small Skills, Big Networks: Marin Mersenne as Mathematical Intelligencer." *History of Science* 51.3(2013): 337–74.

Gwiazda, Jeremy. "Throwing Darts, Time, and the Infinite." *Erkenntnis* 78(2013): 971–75.

Halevy, Avner. "Tails in High Dimensions." *Math Horizons* 20.3(2013): 14–17.

Harel, Guershon. "Classroom-Based Interventions in Mathematics Education." *ZDM Mathematics Education* 45(2013): 483–89.

Heine, George. "Euler and the Flattening of the Earth." *Math Horizons* 21.1(2013): 25–29.

Hickey, Walter. "The Largest Employer of Mathematicians in the U.S.A." *Business Insider* June 10, 2013.

Higgs, Megan D. "Do We Really Need the S-Word?" *American Scientist* 101.1(2013): 6–9.

Hollins, Christopher. "The struggle against Idealism: Soviet Ideology and Mathematics." *Notices of the AMS* 60.11(2013): 1448–58.

Hooper, Martyn. "Richard Price, Bayes' Theorem, and God." *Significance* (2013): 36–39.

Howson, Colin. "Hume's Theorem." *Studies in History and Philosophy of Science* 44(2013): 339–46.

Howson, Geoffrey. "Historical Reflections on the Development of Mathematics Textbooks." *ZDM Mathematics Education* 45(2013): 647–58.

Hutchinson, Martin. "The Mathematical Model Menace." *The Prudent Bear* May 6, 2013.

Jardine, Lisa. "Mary, Queen of Math." *BBC Magazine*, March 8, 2013.

Ji, Lizhen, and Athanase Papadopoulos. "Historical Development of Teichmüller Theory." *Archive for History of Exact Sciences* 67(2013): 119–47.

Johansson, Ingvar. "Constitution as a Relation within Mathematics." *The Monist* 96.1(2013): 87–100.

Kanovei, Vladimir, Mikhail G. Katz, and Thomas Mormann. "Tools, Objects, and Chimeras: Connes on the Role of Hyperreals in Mathematics." *Foundations of Science* 18(2012): 259–96.

Katz, Karin Usadi, and Mikhail G. Katz. "A Burgessian Critique of Nominalistic Tendencies in Contemporary Mathematics and Its Historiography." *Foundations of Science* 17(2012): 51–89.

Katz, Mikhail G., and Eric Leightnam. "Commuting and Noncommuting Infinitesimals." *The American Mathematical Monthly* 120.7(2013): 631–41.

Katz, Mikhail G., and David Sherry. "Leibniz's Infinitesimals." *Erkenntnis* 78(2013): 571–625.

Katz, Mikhail G., David M. Schaps, and Steven Shnider. "Almost Equal: The Method of Adequality from Diophantus to Fermat and Beyond." *Perspectives on Science* 21.3(2013): 283–324.

Kim, Joongol. "What Are Numbers?" *Synthese* 190(2013): 1099–112.

Kimball, Miles, and Noah Smith. "The Myth of 'I'm Bad at Math'" *The Atlantic Online* October 28, 2013.

Klarreich, Erica. "How to Build Impossible Wallpaper." *Simmons Foundation* March 5, 2013.

Ko, Yi-Yin, and Eric J. Knuth. "Validating Proofs and Counterexamples across Content Domains." *The Journal of Mathematical Behavior* 32(2013): 20–35.

Kriegeskorte, Nikolaus, and Rogier A. Kievit. "Representational Geometry: Integrating Cognition, Computation, and the Brain." *Trends in Cognitive Sciences* 17.8(2013): 401–12.

Lambert, Kevin. "A Natural History of Mathematics: George Peacock and the Making of English Algebra." *Isis* 104.2(2013): 278–302.

Landers, Mara G. "Towards a Theory of Mathematics Homework as a Social Practice." *Educational Studies in Mathematics* 84(2013): 371–91.

Landy, David, Noah Silbert, and Aleah Goldin. "Estimating Large Numbers." *Cognitive Science* 37(2013): 775–99.

Lang, Robert J. "A Pajarita Puzzle Cube in Papiroflexia." *Journal of Mathematics and the Arts* 7.1(2013): 1–16.

Lange, Marc. "What Makes a Scientific Explanation Distinctively Mathematical?" *British Journal for the Philosophy of Science* 64(2013): 485–511.

Langlands, Robert P. "Is There Beauty in Mathematical Theories?" In *The Many Faces of Beauty*, edited by Vittorio Hösle. Notre Dame, IN: University of Notre Dame Press, 2013.

Lee, Alexander. "Goals and Scope of the Archimedes Palimpsest Transcriptions." *British Society for the History of Mathematics Bulletin* 28.1(2013): 1–15.

Liljedahl, Peter. "Illumination: An Affective Experience?" *ZDM Mathematics Education* 45(2013): 253–65.

Livio, Mario. "Symmetry: From Perception to the Laws of Nature." In *The Many Faces of Beauty*, edited by Vittorio Hösle. Notre Dame, IN: University of Notre Dame Press, 2013.

Lobato, Joanne, Charles Hobensee, and Bohdan Rhodehamel. "Students' Mathematical Noticing." *Journal for Research in Mathematics Education* 44.5(2013): 809–50.

Luitel, Bar Chandra. "Mathematics as an Im/Pure Knowledge System." *International Journal of Science and Mathematics Education* 11(2013): 65–87.

Ma, Liping. "A Critique of the Structure of U.S. Elementary School Mathematics." *Notices of the AMS* 60(2013): 1282–96.

Maffioli, Cesare S. "A Fruitful Exchange/Conflict: Engineers and Mathematicians in Early Modern Italy." *Annals of Science* 70.2(2013): 197–228.

Marquis, Jean-Pierre. "Mathematical Forms and Forms of Mathematics." *Synthese* 190(2013): 2141–64.

Marshall, Douglas Bertrand. "Galileo's Defense of the Application of Geometry to Physics in the *Dialogue*." *Studies in History and Philosophy of Science* 44(2013): 178–87.

McEvoy, Mark. "Experimental Mathematics, Computers, and the A Priori." *Synthese* 190(2013): 397–412.

Moscati, Ivan. "Were Jevons, Menger, and Walras Really Cardinalists? On the Notion of Measurement in Utility Theory, Psychology, Mathematics, and Other Disciplines, 1870–1910." *History of Political Economy* 45.3(2013): 373–414.

Naylor, Mike. "ABACABA—Amazing Pattern, Amazing Connections." *Math Horizons* 20.3(2013): 24–30.

Núñez, Rafael E. "Enacting Infinity: Bringing Transfinite Cardinals into Being." In *Enaction: Toward a New Paradigm for Cognitive Science*, edited by John Stewart, Olivier Capenne, and Ezequiel A. Di Paolo. Cambridge, MA: MIT Press, 2010.

Núñez, Rafael, and Kensy Cooperrider. "The Tangle of Space and Time in Human Cognition." *Trends in Cognitive Sciences* 17.5(2013): 220–29.

O'Halloran, Kay L. "From Calculation to Computation: Historical Changes in Semiotic Landscape." In *The Routledge Handbook of Multimodal Analysis*, edited by Carey Jewitt. Abingdon, UK: Routledge, 2009.

Obersteiner, Andreas, et al. "The Natural Number Bias and Magnitude Representation in Fraction Comparison by Expert Mathematicians." *Learning and Instruction* 28(2013): 64–72.

Oosterhoff, "From Pious to Polite: Pythagoras in the *Res publica litterarum* of French Renaissance Mathematics." *Journal of the History of Ideas* 74.4(2013): 531–52.

Österholm, Magnus, and Ewa Bergqvist. "What Is So Special about Mathematical Texts?" *ZDM Mathematics Education* 45(2013): 751–63.

Pais, Alexandre. "An Ideology Critique of the Use-Value of Mathematics." *Educational Studies in Mathematics* 84(2013): 15–34.

Pengelley, David. "Quick, Does 23/67 Equal 33/97? A Mathematician's Secret from Euclid to Today." *The American Mathematical Monthly* 120.10(2013): 867–76.

Pesic, Peter. "Euler's Musical Mathematics." *Mathematical Intelligencer* 35.2(2013): 35–43.

Pittalis, Marios, and Constantinos Cristos. "Coding and Decoding Representations of 3D Shapes." *The Journal of Mathematical Behavior* 32(2013): 673–89.

Polster, Burkard, and Marty Ross. "The Golden Ratio Must Die!" *The Age*, Australia, March 4, 2013.

Price, Gavin. "Dyscalculia: Characteristics, Causes, and Treatments." *Numeracy* 6.1(2013): 1–16.

Priestley, W. M. "Analogy, Ambiguity and Humanistic Mathematics." *Journal of Humanistic Mathematics* 3.1(2013): 115–35.

Radin, Charles. "A Revolutionary Material." *Notices of the AMS* 60.3(2013): 310–15.

Rågstedt, Mikael. "Kepler and the *Rudolphine Tables*." *Bulletin of the AMS* 50.4(2013): 629–39.

Rips, Lance J., Jennifer Asmuth, and Amber Bloomfield. "Can Statistical Learning Bootstrap the Integers?" *Cognition* 128(2013): 320–30.

Rizza, Davide. "The Applicability of Mathematics: Beyond Mapping Accounts." *Philosophy of Science* 80.3(2013): 398–412.

Roth, Wolff-Michael. "Contradictions and Uncertainty in Scientists' Mathematical Modeling and Interpretation of Data." *The Journal of Mathematical Behavior* 32(2013): 593–612.

Rowe, David E. "Who Linked Hegel's Philosophy with the History of Mathematics?" *Mathematical Intelligencer* 35.1(2013): 38–41 and 35.4(2013): 51–55.

Rowlett, Peter. "Developing a Healthy Skepticism about Technology in Mathematics Education." *Journal of Humanistic Mathematics* 3.1(2013): 136–49.

Sandefur, James, and John Mason. "Circular Inclusion." *The College Mathematics Journal* 44.3(2013): 193–201.

Sandefur, J., J. Mason, G. J. Stylianides, and A. Watson. "Generating and Using Examples in the Proving Process." *Educational Studies in Mathematics* 84(2013): 323–40.

Schlimm, Dirk. "Axioms in Mathematical Practice." *Philosophia Mathematica* 21(2013): 37–92.

Schoenfeld, Alan. "Classroom Observations in Theory and Practice." *ZDM Mathematics Education* 45(2013): 407–21.

Schoenfeld, Alan. "On Forests, Trees, Elephants, and Classrooms: A Brief for the Study of Learning Ecologies." *ZDM Mathematics Education* 45(2013): 491–95.

Schoenfeld, Alan. "Reflections on Problem Solving Theory and Practice." *The Mathematical Enthusiast* 10.1–2(2013): 9–34.

Schoenfeld, Alan, and Jeremy Kilpatrick. "A U.S. Perspective on the Implementation of Inquiry-Based Learning in Mathematics." *ZDM Mathematics Education* 45(2013): 901–909.

Schulman, Steven M. "According to Davis: Connecting Principles and Practices." *The Journal of Mathematical Behavior* 32(2013): 230–42.

Schuster, Peter. "A Silent Revolution in Mathematics." *Complexity* 18(2013): 7–10.

Segerman, Henry. "How to Print a Hypercube." *Math Horizons* 20.3(2013): 5–9.

Seife, Charles. "An Open Letter to My Former NSA Colleagues." *Slate* August 22, 2013.

Shipman, Barbara A. "Mathematical Versus English Meaning in Implication and Disjunction." *Teaching Mathematics and Its Applications* 32.1(2013): 38–46.

Slater, Hartley. "Symbols and Their Meaning in Analysis." *Logique et Analyse* 56.222(2013): 211–25.

Smith, Kim, Elaine Hodson, and Tony Brown. "The Discursive Production of Classroom Mathematics." *Mathematics Education Research Journal* 25(2013): 379–97.

Soifer, Alexander. "What 'Problem Solving' Ought to Mean." *Mathematical Competitions* 26.1(2013): 8–22.

Steen, Lynn Arthur. "Can We Make School Mathematics Work for All?" *Notices of the AMS* 60.11(2013): 1466–70.

Stewart, Ian. "The Power and Glory of Mathematics." *The New Statesman* May 21, 2013.

Stillwell, John. "Poincaré and the Early History of 3-Manifolds." *Bulletin of the AMS* 49.4(2013): 555–76.

Suri, Manil. "How to Fall in Love with Math." *The New York Times* September 15, 2013.

Tallant, Jonathan. "Pretense, Mathematics, and Cognitive Neuroscience." *British Journal for the Philosophy of Science* 64(2013): 817–35.

Tucker, Alan. "The History of the Undergraduate Program in Mathematics in the United States." *The American Mathematical Monthly* 120.8(2013): 698–705.

Tymoczko, Dmitri. "Geometry and the Quest for Theoretical Generality." *Journal of Mathematics and Music* 7.2(2013): 127–44.

Vamvakoussi, Xenia, et al. "The Framework Theory Approach Applied to Mathematics Learning." In *International Handbook of Research on Conceptual Change*, edited by Stella Vosniadou, pp. 305–321. New York: Routledge, 2013.

van Oers, Bert. "Challenges in the Innovation of Mathematics Education for Young Children." *Educational Studies in Mathematics* 84(2013): 267–72.

Vitanyi, Paul M. B. "Tolstoy's Mathematics in *War and Peace*." *The Mathematical Intelligencer* 35.1(2013): 71–75.

Vogel, Rose. "Mathematical Situations of Play and Exploration." *Educational Studies in Mathematics* 84(2013): 209–25.

Vukovic, Rose K., and Nonie K. Lesaux. "The Relationship between Linguistic Skills and Arithmetic Knowledge." *Learning and Individual Differences* 23.1(2013): 87–91.

Wagner, Roy. "A Historically and Philosophically Informed Approach to Mathematical Metaphors." *International Studies in the Philosophy of Science* 27.2(2013): 109–35.

Ward, Mark. "How the Modern World Depends on Encryption." *BBC News* October 25, 2013.

Williams, Travis D. "The Dialogue of Early Modern Mathematical Subjectivity." *Configurations* 21.1(2013): 53–84.

Wilson, E. O. "Great Scientist ≠ Good at Math." *The Wall Street Journal* April 5, 2003.

Zabell, Sandy. "Paul Meier on Legal Consulting." *The American Statistician* 67.1(2013): 18–21.

Zelcer, Mark. "Against Mathematical Explanation." *Journal for General Philosophy of Science* 44(2013): 173–92.

Notable Journal Issues

The following special issues of 2013 journals (or group of articles on a featured theme in general-audience magazines) is likely to appeal to the reader interested in nontechnical writings and topics concerning mathematics. This list is selective.

"Summer School" *The American Mathematical Monthly* 120.3(2013).

"The Mathematics of Planet Earth" *The College Mathematics Journal* 44.5(2013).

"Problem Posing in Mathematics Teaching and Learning" *Educational Studies in Mathematics* 83.1(2013).

"Marc Barbut (1928–2011)" *Electronic Journal for History of Probability and Statistics* 9(2013).

"The Teaching and Learning of Calculus" *International Journal of Mathematical Education in Science and Technology* 44.5(2013).

"Quantitative Skills in Science" *International Journal of Mathematical Education in Science and Technology* 44.6(2013).

"Frege's Philosophy of Mathematics and Language" *History and Philosophy of Logic* 34.3(2013).

"Equity [in mathematics education]" *Journal for Research in Mathematics Education* 44.1(2013).

"Teaching Abstract Algebra for Understanding" *Journal of Mathematical Behavior* 32.4(2013).

"Mathematical Theories of Voice Leading" *Journal of Mathematics and Music* 7.2(2013).

"Calibrating Calibration: Creating Conceptual Clarity to Guide Measurement and Calculation" *Learning and Instruction* 24.2(2013).

"The Language(s) of Mathematics" *Logique & Analyse* 56.221(2013).

"Language and Mathematics Education" *Mathematics Education Research Journal* 25.3(2013).

"International Perspectives on Problem Solving Research in Mathematics Education" *The Mathematics Enthusiast* 10.1–2(2013).

"Financial Literacy" *Numeracy* 6.2(2013).

"Logicism Today" *Philosophia Mathematica* 21.2(2013).

"Capstone Courses" *PRIMUS* 23.4(2013).

"Service-Learning in Mathematics" *PRIMUS* 23.6(2013).

"Undergraduate Research in the Mathematical Sciences" *PRIMUS* 23.9(2013).

"Polyhedra" *Symmetry* 5.1(2013).

"Mathematics Applied to the Climate System" *Philosophical Transactions of the Royal Society, Series A, Mathematical, Physical and Engineering Sciences* 371.1991(2013).

"The African American Mathematics Teacher" *Teachers College Record* 115.2(2013).

"Creativity and Mathematics Education" *Zentralblatt für Didaktik der Mathematik* 45.2(2013).

"Theoretical Frameworks in Research on and with Mathematics Teachers" *Zentralblatt für Didaktik der Mathematik* 45.4(2013).

Acknowledgments

This volume would not exist without the cooperation of its contributors. Thank you to all—and thanks to the original publishers of the material, including the owners of copyrighted pictures. Also thanks to the anonymous reviewers who rated and commented on the extended collection of writings I initially proposed for evaluation. Yet I am entirely responsible for the flaws you might find with the content of this book.

Thanks to Vickie Kearn for her solicitude at all stages and for advice at key decision moments. Also at Princeton, thanks to Quinn Fusting for solving the copyright issues and to Nathan Carr for overseeing the production. Paula Bérard copyedited the manuscript; thank you.

On the academic side, I am on leave this year, missing teaching and the income it used to bring; this threw me back into survival mood, an intermittently perennial condition for which I achieved notorious qualifications. I still thank Professors David W. Henderson, John W. Sipple, and Steven Strogatz, for not giving up on me. I always felt that gaining academic degrees is a nuisance incongruous with my intrinsic motivation to learn and with my indifference to officially sanctioned credentials.

Thanks to Robert H. Davis of Columbia University and to Pedro Arroyo of Cornell for helping me bring my massive personal library to the United States, over several seas and an ocean. This made it possible for me to donate thousands of foreign books and recordings, most of them rare or unique in this country, to the Cornell Library—a pittance of gratitude for the many services it provides to me.

Fangfang has been steady in doing the right thing at the right time, a skill I would do well to learn myself. Thanks!

Lastly, to my daughter Ioana. Nothing quite matches the feelings of a father (a full-time caring father) who sees his child dragged gratuitously into the balderdash-hosting offices that mushroom around family courts. I was in that situation nine years ago, just when I attempted to start this series of anthologies. Later, when I read *On Bullshit* by

Harry Frankfurt, it dawned on me that my daughter and I were at the receiving end of professional practices that enable, endorse, covet, and thrive on bullshit. My daughter faced the ordeal confused, yet bravely (as much as a three-year-old could face things bravely), but we were no match for the coteries of unaccountable attorneys and "psycho" people who hosted and minded the show. For remaining a steadfast, loving daughter throughout that tormenting time and ever since—despite the relentless pressure on her to do the opposite—I thank her and I dedicate this volume to her.

Credits